DECISION MAKING IN POLICING

Forensic sciences

DECISION MAKING IN POLICING

OPERATIONS AND MANAGEMENT

Pierre Aepli, Olivier Ribaux, Everett Summerfield

EPFL Press
A Swiss academic publisher distributed by CRC Press

CRC Press
Taylor & Francis Group

Taylor and Francis Group, LLC
6000 Broken Sound Parkway, NW, Suite 300,
Boca Raton, FL 33487

Distribution and Customer Service
orders@crcpress.com

www.crcpress.com

Library of Congress Cataloging-in-Publication Data
A catalog record for this book is available from the Library of Congress.

This book is published under the editorial direction of
Professor Pierre Margot (University of Lausanne).

EPFL Press

is an imprint owned by Presses polytechniques et universitaires romandes, a Swiss academic publishing company whose main purpose is to publish the teaching and research works of the Ecole polytechnique fédérale de Lausanne (EPFL) and other universities and institutions of higher learning.

Presses polytechniques et universitaires romandes
EPFL – Rolex Learning Center
Post office box 119
CH-1015 Lausanne, Switzerland
E-mail: ppur@epfl.ch
Phone: 021 / 693 21 30
Fax: 021 / 693 40 27

www.epflpress.org

ISBN 978-2-940222-54-4 (EPFL Press)
ISBN 978-1-4398-7327-4 (CRC Press)

Printed in France by Présence Graphique – N° d'imprimeur: 111038402

To Guillaume, Anthea, Oriane and Olivia;
To Julene, Coady, Anne Marie, Annette and Tansi;
To the memory of François, Fred and Jean-Daniel.

Foreword

The best known fictional detective of all time, Sherlock Holmes, was famous for saying that the deductive process was "elementary." In my 40 years as a police officer I can't say I have found that to be true. Making good and right decisions, both in response to problems and in order to determine a best course of action, was probably the most difficult thing I have had to do.

Over the course of my career I made thousands of decisions, from the day to day responses of a beat cop, to the more analytical work of a junior officer in commercial crime, and into a variety of managerial roles, including the challenging six years I spent as Commissioner of the Royal Canadian Mounted Police overseeing a full range of operations from community policing to national security and everything in between.

As I read through this remarkably thorough document it struck me how much it would have helped me to have had access to such a superb reference table and training document during those years. I can think of a number of ways I'd have used it; to refer to when reflecting on making, revising or responding to a decision. Or, as an analytical guide for use in that common exercise we all engage in – the post mortem after the closure of an investigation or tactical operation. Finally, this resource would have provided me with the tools to transmit critically important skills and knowledge on to those I had responsibility to train, or manage.

However, as I spent more time with the document, thinking about what I wanted to say in this Foreword, I found myself also pondering the strategic importance of it – beyond its obvious potential to aid and abet policing operations.

Over the past ten years or so my work and thinking has begun to focus more on systemic and even philosophical matters, than the operational ones of my past. At the risk of oversimplification, my conclusion about how law enforcement needs to evolve centers around two key concepts; *intelligence-led approaches* and *integrated policing*. As I read and re-read the material presented in this book, a kind of circular truth bubbled up that gets to the very heart of what we need to leverage this evolution:

- Good decision making enables police officers and organizations to take raw information and transform it into intelligence, through a process of assessment, analysis, testing, evaluation…
- In turn, effective intelligence guides good decision making by officers and organizations – at the operational, managerial and strategic levels.

Understanding all the elements of effective decision making, and learning how to harness and utilize them, will provide the needed bedrock for a widespread adoption of intelligence based approaches – at all levels and across jurisdictions in law enforcement. And a comprehensive adopting of intelligence-led policing will fundamentally change the way decisions are made, resources are allocated, and criminal activity is combated, in all our various contexts.

What is particularly exciting to me about the decision-making paradigm presented here is its universality. It provides a generic enough process guide that it could be used as effectively in a large urban center of a first world country, as in the smallest of police forces in a rural setting somewhere in the third world. This is exactly what we need to make the second of my "platforms for 21st century law enforcement" a reality.

Integrated operations, or strategic direction setting, brings leaders and officers from wildly divergent realities together. Nothing could be more important than for the law enforcement community to position itself to collectively respond to the "new" realities of terrorism, globally connected organized crime, the trans-border challenges of environmental, bio-hazard or cyber crime. Nothing is ever more challenging. Working *alongside* people from other countries or even the police force from the next town is one thing – working as one unit and creating the seamless and conjoined process that an integrated approach demands, is an altogether different kettle of fish. Having a shared point of reference on decision making process will go a long way to helping us overcome the barriers we have all had to deal with, as we move toward truly integrated police work.

When I was asked to write the Foreword to *Decision Making in Policing* I figured I would say something nice about a well written training resource on an important element of police development. What I hadn't thought through, to that point, was how a concrete tool such as this could in fact leverage the more existential vision for the future of law enforcement, that many of us have been talking about for quite a while now.

I believe *Decision Making in Policing* has the potential to become a seminal reference for law enforcement that changes the standards around how we think about, and enact decision making methodology. Most importantly, it provides the tactical framework to operationalize intelligence-led and integrated policing around the world. Which in the end is where we need to go. So that each of us, who have made an oath to serve our countries and our citizens, can be part of a shared effort to be the most effective agents for safety and security possible.

Giuliano Zaccardelli
August 27, 2010

Giuliano Zaccardelli is the Director of Strategic Planning for INTERPOL. For almost thirty seven years he was a member of the Royal Canadian Mounted Police (RCMP). From 2000 until his retirement from the RCMP in 2006 he was the head of the organization. For many years he has been a champion and outspoken proponent of intelligence-led and integrated policing at the local, national and international levels as a means to deal with the threat of the globalization of organized crime, public safety and security issues.

Preface

Some years ago, Pierre Aepli, in his role as consultant for the Geneva Centre for the Democratic Control of Armed Forces (DCAF), agreed to accept the responsibility of developing the working group *Intelligence, Risk Analysis and Investigation* of DCAF's Border Security Programme in Southeast Europe. Its goal was to assist the governments of South-East Europe (SEE) to establish reliable and efficient border security systems; thus the DCAF and its partners developed a program aimed at providing assistance, ranging from national-capacity building to the development of regional operational cooperation mechanisms. The *Intelligence Working Group* aimed at setting up structures, methods and procedures to improve decision-making and cooperation processes through the integration of intelligence analysis.

Olivier Ribaux, who had contributed to the introduction of analysis in the police force led by Pierre Aepli, and who was teaching crime and forensic intelligence at the University of Lausanne, was part of the working group. Pierre and Olivier, together with Inge Lindsaar and Marko Saareks, two officers of the border guard of Estonia and Finland, respectively, developed and conducted a series of three seminars over the course of several years.

The experiences gained during the seminars – specifically the preparation, implementation and debriefing phases of the activity – led us to identify the clear need to better integrate intelligence into the decision-making process. Pierre and Olivier decided to move beyond the seminar to take advantage of all their experience, related to the coursework as well as to their professional activities, to write a book to address the issue. The first approach was to focus on the role of intelligence in decision-making, but it immediately became clear that the real subject was the decision-process itself. Intelligence remained a capital factor in this process, but the authors saw the benefit of emphasizing the different phases of the process and to place it within its organizational environment; hence the developments tackling the issues of strategy, structures, and leadership.

The experience of our third author Everett Summerfield, a chief superintendant of the Royal Canadian Mounted Police seconded to DCAF, provided a new perspective that proved beneficial during the preparation of the book. Friendship, professionalism, excellent coordination and absolute respect of the deadlines have been the factors of success of the editing of the book.

Decision Making in Policing has been structured on three levels to optimize its utility to the reader. Theoretical elements are provided in order to introduce fundamental aspects and existing approaches of decision-making in policing. To render the reading less abstract, case studies are developed to illustrate the concepts through the description of practical situations. Finally, the book makes extensive use of tables, which aim to provide a synthesis between the authors' personal experiences and the interpretation of the theory. In addition, they should be used as checklists that support decision makers in elaborating their own procedures.

The authors address *Decision Making in Policing* to managers of the police at all levels, where it will prove useful during their daily tasks; in addition, it is intended that the book bring an fresh and informative perspective to all those who are interested in understanding, organizing and improving law-enforcement activities.

ACKNOWLEDGEMENTS

We would like to acknowledge the contributions of a number of individuals who have contributed to the realization of this book, from its inception to this end product.

A profound thanks to DCAF and the Ecole des Sciences Criminelles (ESC), University of Lausanne, as the institutions that drove this collaborative work; Darko Stancic, head of operations at DCAF who supported this project and Oovel Andrus, head of the Border Security Program at DCAF for his valued input.

The Police cantonale vaudoise has hosted many of the experiments mentioned in this book.

In a project such as this, there are various participants and drivers who have invested time and resources. Special thanks to Inge Lindsaar, Marko Saareks and all other contributors and participants of DCAF seminars organized within the framework of the programme, *Border Guard Security Intelligence in the Balkans (2006-2009)*. We greatly appreciate the work of Yury Korobovsky, Christopher Thornton (DCAF) as well as Quentin Rossy and Sylvain Ioset (ESC) for their contributions to the various figures in the book. Our heartfelt thanks to Daniela Abeztout Neuenschwander (DCAF) for the management of the different versions of the manuscript and the reviewing of the bibliography and to lt-Colonel Alain Bergonzoli who provided us with useful documentation. Further thanks to Alexis Kalagas who proficiently reviewed the text as well as to Alero Okorodudu and Nicholas Margot who participated in reviewing previous versions of the manuscript. Gratitude to Giulianni Zaccardelli for the forward.

We are immensely grateful for all your contributions to this project.

About the Authors

The three authors of this book have each acquired significant experience in police matters and have all been active – at different levels (national or international) and in different areas (police, military, business and academia) – in using and developing information and decision-making systems. The intent of this collaboration is to contribute to the improved awareness of readers of this book in the importance of establishing an efficient decision-making process within police organizations.

Pierre Aepli

Pierre Aepli worked for 16 years for a private company as managing director in charge of several foreign subsidiaries. Following this, he led the 1000-member-strong police force of the canton of Vaud from 1982 to 2002, and for several years he chaired the Swiss Police Chiefs Conference, which is the coordinating body of the Swiss police forces. He initiated and conducted many police reforms at the cantonal and national levels in Switzerland. Since his retirement he has chaired, in 2003, the Swiss Directing Committee responsible for the organisation of the G8 summit and has performed various international missions in the Balkans, the former Southern Soviet Republics, Indonesia and Palestine for Switzerland, the Council of Europe and for DCAF. He has published several articles on security issues.

Pierre holds a degree in political science, a diploma from IMD and is a graduate of the FBI National Academy in Quantico, Virginia. He has taught a course on police management for the University of Lausanne at the Ecole des Sciences Criminelles. In the Swiss Army he was a colonel in command of an infantry regiment. He is married and has two children and four grandchildren. He lives in Rolle, Switzerland.

Olivier Ribaux

Olivier Ribaux is a Professor at the Ecole des Sciences Criminelles (ESC) at the University of Lausanne. With a background as a mathematician, he previously worked in software development for private companies and a Swiss police force. After a Ph.D. involving research on crime analysis and forensic intelligence, Olivier worked for eight years at the police of the canton de Vaud. During this period, he developed crime intelligence methods and tools within the framework of an agreement between several police forces in Switzerland. He has also worked as a consultant for a number of international organisations, including DCAF. He has authored many peer-reviewed scientific papers and book chapters on policing, crime intelligence and forensic science.

Everett Summerfield

Chief Superintendent Everett Summerfield is a serving officer with Canada's Royal Canadian Mounted Police. In his 36 years with the RCMP he has served in a range of duties and commands at various locations throughout Canada and abroad with the United Nations in Kosovo (1999-2000) and the European Union Police Mission in Afghanistan (2008-2009). Prior to joining the RCMP in 1974 he served six years within the Canadian Armed Forces as Military Police and with Canadian Airborne Regiment at various locations in Canada and Cyprus (1974). He has a diverse background in Operational Policing, Operational Support, Administration and Senior Executive Management. Chief Superintendent Summerfield is currently seconded to the Geneva Centre of the Democratic Control of Armed Forces as the Manager of the Police Program, which provides strategic direction in leadership and management of police services and governance. He is married to Julene and they have four children, Coady, Anne Marie, Annette and Tansi.

About DCAF

The Geneva Centre for the Democratic Control of Armed Forces (DCAF) is an international foundation whose mission is to assist the international community in pursuing good governance and reform of the security sector.

To this end, the Centre develops and promotes appropriate norms at the international and national levels, determines good practices and relevant policy recommendations for effective governance of the security sector, and provides in-country advisory support and practical assistance programmes to all interested actors.

DCAF's principal fields of expertise include: capacity building in security sector reform (SSR) for civil society and the media, gender and SSR, intelligence oversight and intelligence accountability, ombuds institutions for security sector personnel, parliamentary oversight for the security sector, police and border police management and cooperation, private military and security company regulation, SSR assessment, SSR monitoring and evaluation, SSR programme design and guidance, SSR strategy development, and SSR training.

Detailed information is available at www.dcaf.ch

DCAF
a centre for security, development and the rule of law

Geneva Centre for the Democratic Control of Armed Forces (DCAF)
Rue de Chantepoulet 11, PO Box 1360, CH-1211 Geneva 1, Switzerland
Tel: +41 22 741 77 00; e-mail: info@dcaf.ch

Table of Contents

Introduction

Every police officer has to make decisions on a daily basis. A decision entails a choice between different approaches to solving a problem. The consequences of these decisions may be trivial or may have repercussions on the lives of others. Frequently, if not always, time, information and analytical capabilities are lacking, resulting in rational decisions being supplanted by intuitive reactions with unforeseen consequences. While there is no way to ensure that in each case a sound decision will always be made, it is possible nonetheless to reduce the risks of poor outcomes by selecting the right people and preparing them through teaching and training. To such a degree, a clear decision-making process should not only reduce the risks of poor decisions, it should also offer a useful method to follow in times of crisis, as a means to restore a degree of order amidst otherwise chaotic conditions. In such contexts, the use of a specific methodology for decision-making represents a considerable advantage, and this book is designed for that purpose.

The function of this book is intended to inform various aspects that are involved in the decision-making process of policing. Firstly, it is the aim of this book to offer a comprehensive decision-making methodology and provide elements intended to assist those required to make decisions in a law enforcement context. The contents of this book are not constrained by the level of hierarchy at which an individual operates; the elements of the decision-making process remain the same irrespective of hierarchy. Accordingly, the book will focus systematically on different phases of the decision-making process. This will be expounded on not only by examining in detail each of the phases involved, but also by considering the broader context within which the process takes place, notably the operational context. Furthermore, the book will employ real life examples in the discussion of issues such as operational structures, tactics, and command, allowing it to serve as a tool for both academics and practitioners. Practices and examples taken from the military and the business world will be used to refer to the breadth of applicability of the decision-making methodology. Many of these practices and methods, particularly in the management[1] of operations and administrative affairs, have been integrated into police framework.

---

1 Peter Drucker, *Management, Tasks, Responsibilities, Practices*, Management Editions (Europe) Dirmas S.A., Geneva, 1974.

As mentioned, a primary task of this book is to outline the possible improvement of the decision making process; but before considering the prospects of improvement, certain conditions need to be met:

- A methodology is developed and this methodology complements the entirety of the decision-making process.
- A necessary coherence links the ends, the ways and the means in which decisions are made and implemented.
- The duties and responsibilities between different actors are clearly defined, as well as their roles at different phases of the process.
- Adequate structures, tactics, leadership styles and command processes support the decision-making process.
- Acknowledgment of the key role of information at all phases of the process.
- Identification of biases and hindrances in the decision-making process.

Correspondingly, this book will emphasize the fundamental function of "information" throughout the entire decision-making process. Though the specific understanding of "information" adopted here will be dealt with in detail later in the book, it is important to emphasize at the outset that while intelligence-led policing has become very popular, the full value of this approach can only be realized if intelligence is integrated holistically throughout the decision, implementation and evaluation processes.[2] For this reason, intelligence-led policing is not a *stand-alone* concept that is in opposition to other models of policing, such as *problem-oriented policing* or *community-oriented policing*, but rather they are mutually enforcing. For instance, intelligence-led policing is a top-down model while community oriented policing is considered a bottom-up model. The integration of these different models – conceived as a means of attaining greater efficiency – must not only be achieved through the implementation of intelligence systems within police forces, but must also be translated into organizational measures, procedures, leadership styles, and training programs.

How then does this book contribute to current work in this field? Indeed, the added value of this book is derived from its integrated and systemic approach, situating the decision-making process within a larger context, as well as identifying the actors involved. The complexity of each specific process will naturally depend upon the type of decision and the level at which it is taken, but the basic methodology will in most cases remain the same.

Each chapter follows a consistent structure:

- A short introduction;
- A theoretical section;
- Practical applications with methods, tables, check lists and examples;
- A summary of key points and factors that are most likely to contribute to risks or successes during the implementation phase.

---

2 At this point, *information* and *intelligence* are used interchangeably as data interpreted, presented, and understood specifically in order to assist police in making decisions according to the various aspects described in this book. In Chapter 2 (Information), we will discuss usual distinctions made between these terms.

Chapter 1

Fundamentals

Objectives

This chapter presents a general framework for the different issues to be further developed in subsequent chapters. A systemic approach has been adopted on the basis that police organizations – like any other type of organization – operate using systems. The decision-making process therefore represents one of the mechanisms constituting the broader system.

1.1 STRUCTURE

The first part of the chapter is devoted to the presentation of the systemic approach, explaining what a system is and how it functions. The chapter goes on to describe the decision-making process, with a particular emphasis on the elements required, as well as their integration in achieving the end result. Subsequently, the chapter presents different types of decisions and examines the role of information in shaping each decision, while also addressing the potential risks and problems that can distort the decision-making process. The chapter concludes by identifying the key factors contributing to success or failure, which should be taken into account by decision-makers.

1.2 THE SYSTEMIC APPROACH

1.2.1 What is a system?

A system can be defined as an ensemble of elements and actors, interacting mutually and with their environment, organized to fulfill one or several objectives.

The concept and functioning of a system

The concept of a system consists of the following elements (Figure 1.1):

- The systemic *environment*, which can be divided into:
  - a transnational environment;
  - a general environment conceived of as the immediate environment with which actors within the system directly interact with;
  - a transactional environment within which the different elements of the system (its sub-systems) interact.
- The *inputs* that arise from both the external and internal environment of the system: threats, needs, demands, expectations, support, resources, information, energy.
- The *components* of the system – its constitutive actors in social and human systems – and their *interactions*.
- The *mechanisms* that enable the system to work, learn and adapt.
- The *outputs* (services, products) that are generated by the system and which must satisfy the threats, needs, demands and expectations faced by the system.

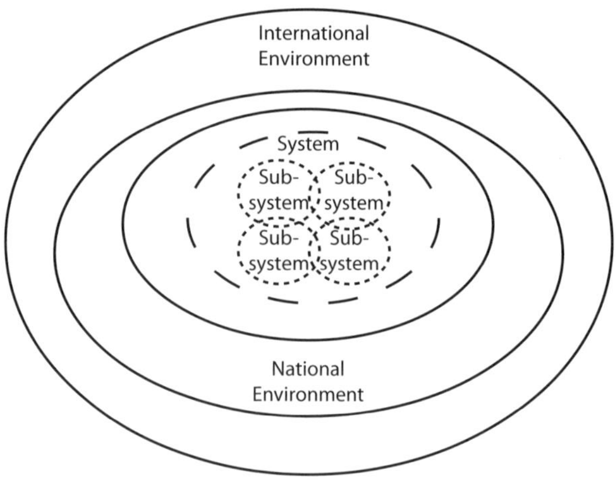

Figure 1.1 A system and its elements.

Characteristics of a system

Figure 1.2 depicts how a system functions.

A system can be closed or open

A system is closed if actors within it do not interact with an external environment. This is generally not the case of social or human systems, which are open systems and continually engage with other systems. Such open systems may be seen as sub-systems of larger systems. A system is constituted of different parts (*e.g.* individuals, groups, organizations or institutions). This means when studying a system

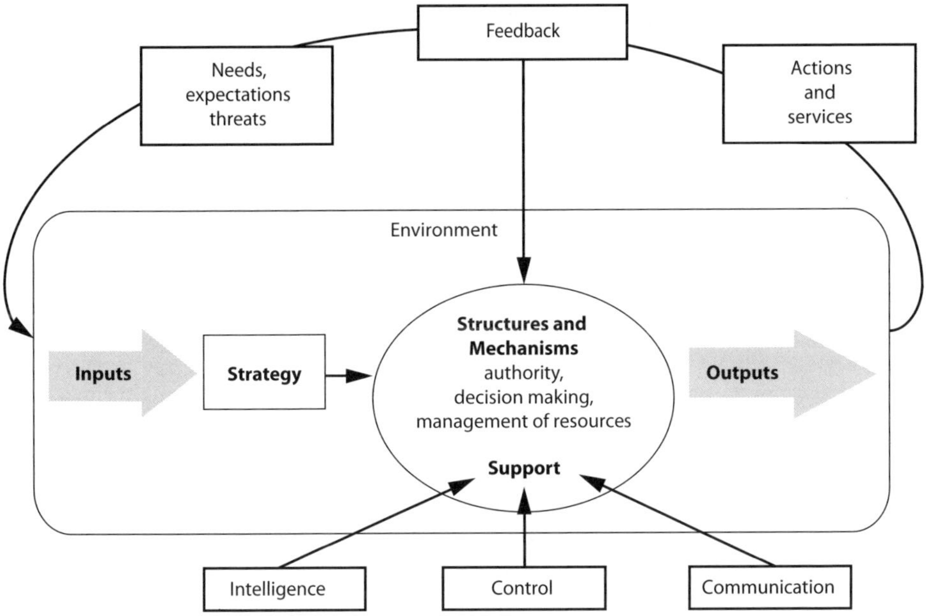

Figure 1.2 A system receives inputs from the broader environment and transforms these into outputs via various mechanisms and procedures. The actions delivered are compared – evaluated – with the original needs and expectations (inputs), in order that where necessary corrections or adjustments can be made.

one has to look at its components, their structures, their relations within the system and with outside systems. One also has to analyze the system's goals or ends, how the system is organized (*e.g.* hierarchically), and the mechanisms and processes involved in the production and delivery of systemic outputs.

Internal interactions within a system and with external systems

It is important to identify interactions within a system and with external systems, as they influence the efficiency and the economy of a system.

Interactions take place at four levels:

- Level 1: Between a system and a supra system. (*e.g. between the security system and the political environment*).
- Level 2: Between a system and its subsystems (*e.g. in the relations between the security policy and the missions given to the army, the police, and the border guards*).
- Level 3: Between subsystems (*e.g. in the cooperation between police and border guards*).
- Level 4: within subsystems, (*e.g. reforms done at the level of an actor*).

These relations can differ in nature: informal (personal links) versus formalized (shaped by procedures). Likewise, they can be distorted by various factors of human, technological, methodological, or organizational natures.

The security system is shown in the Table 1.3.

Table 1.3 A description of the secutity system.

Levels	Relations	Types of problems	Actions
System and suprasystem	Two-way relationship between system and supra-system.	Identification of changes within the broader environment and the ensuing impacts.	Adjustment of national security policy (consequences at both the national and international level).
System and subsystems	Two-way relationship between system and sub-system.	Consequences for the subsystems linked to the changes occurring at the system level. Consequences to be taken into account in the system because of the limits of the resources and capabilities of the subsystems.	Possible adjustments in the missions, roles and resources of the actors of the security system. Possible adjustments to national security policy linked to the constraints posed by the subsystems.
Between subsystems	Horizontally within the actors of the security system.	Consequences for the nature and extent of cooperation between subsystems.	Possible adjustments in the procedures and mechanisms of cooperation between the subsystems.
Within subsystems	Between the different parts (units) of a subsystem.	Consequences within the subsystem for strategy, structures and procedures in order to adapt to the changes of the environment.	Reforms to structures in relation to emerging environmental changes or relations between various actors. Necessary measures to be taken by concerned subsystems.

The limits of a system

It is necessary to differentiate a system from the broader environment. Tracing such limits allows one to separate those who belong to a system from those who do not. If limits are easy to identify in mechanical systems, they are much more complex to define in human systems. How can one do so?

The limits of a system are not rigid, as the evolution of the broader environment influences the threats, needs or expectations that actors within the system face. New actors can emerge, or mechanisms may need to be altered in order to respond to specific developments. Each system must *periodically* adapt its goals, outputs, and ways of functioning *to the evolution of its environment.* Private security companies, for example, have expanded into new and sensitive activities in the field of security systems with implications to existing regulatory regimes and roles of other players. For instance, police may transfer specific surveillance tasks to private contractors; in many cases, legislation must be amended. Similarly, structures and mechanisms may also have to be adapted: an example can be found here in the

Case Study: The National Security System

It is useful to begin with the case of a national security system. Why does such a system exist? The role of a national security system is to ensure the defense of a national territory and the safety of the relevant society, to protect its citizens and their properties, to maintain the public order, to prevent crime (if necessary to identify, arrest and uphold offenders to the justice system), and to defend the state and its institutions. Given the broad definition of this goal, the perimeter of the system will vary. For instance, the notion of human security, encompassing, *inter alia*, social, economic or health security, is conceptually more extensive than the notion of national security focused on state-centered issues of external and internal security. When the goal has been defined, the next step is to identify the actors who contribute to its achievement. The limits of relevant sectors can then be aligned according to these actors. If security is denoted as a policy aimed at achieving the external and internal security of a country, one can classify the police, the army, border guards, the intelligence services and private security companies as the relevant actors. If a broader definition is preferred, one would have to include additional actors dealing with other policy dimensions such as health and social services.

development of communications, linked to the growing importance of the media and public opinion in all security issues.

Interaction and cooperation between actors can also be affected, as well as their placement within a system. In Europe, the implementation of the Schengen agreements[1] has led in many countries to the merger of the former border guards with the police. The perceived growth of crime and terrorism has also fostered cooperation between armies and police forces in several countries: this is especially the case in France, with the Plan *Vigipirate*, and in Italy, where the army has been mobilized in support of the police in many anti-mafia operations. The limitations and challenges to a system are fluid due to the ever-changing influence of the environment. Vigilance is required to adapt to the changing environment ensuring efficiency and effectiveness.[2]

1 Belgium, France, Germany, Luxembourg and the Netherlands agreed, via the Schengen Agreement (signed 14 June 1985), to gradually remove controls at their common borders and introduce freedom of movement for all nationals of signatory Member States, other Member States or third countries. Since 1999, this has formed part of the institutional and legal framework of the European Union. The Schengen agreements have been extended over time to 15 old Member States: Two non-Community countries – Iceland and Norway – have also belonged to the Schengen area since 1996, though they have only a limited role in decision-making. Switzerland also participates in the Schengen area since 2008 under the same status as Norway and Iceland.

2 "Effectiveness is the foundation of success-efficiency is a minimum condition for survival after success has been achieved. Efficiency is concerned with doing things right; effectiveness is doing the right things" (Peter Drucker, *Management, Tasks, Responsibilities*, Practices, Management Editions (Europe) Dirmas S.A., Geneva, 1974, p. 45).

The structures of a system

Designing the structures of a system, including the component elements, is a primary issue. The basic principle "structures follow strategy" does not always apply as constraints determined by the political structure or history of a country, can strongly limit the types of reforms achievable within a police organization. For example, all attempts to merge local police forces in a federal system – like in the United States or Switzerland – are very difficult, as the decentralization of political authority means state (or canton) level governments will likely oppose any perceived centralization of powers. Therefore, one of the main concerns in a security system is to find the best balance between which elements should be centralized and which elements should remain decentralized; the answer to this question will vary on the basis of specific national contexts.

The mechanisms of a system

One of the requirements of every system is functional efficiency. To achieve this aim, a certain number of conditions must be fulfilled. The system's goals must be defined and translated into clear objectives, and systemic structures and work processes must be adequate to meet these ends with the means available. The Table 1.4 presents some mechanisms that are necessary for the operation of a system.

We summarize here the challenges of the mechanisms of a system:

- Managing relations with the environment from which inputs are received and to which outputs are supplied.
- Being organized in an efficient way.
- Ensuring its survival through an ability to adapt to the prevailing environment and subsequent developments (this is a key difference between private and public systems; the failure of the former to adapt to their environment leads to their disappearance and the inadequacy of the latter systems simply renders them more or less inefficient).

The efficiency of a system

The efficiency of a system can be measured in three ways. Firstly, by the conformity of the outputs produced by the system with the threats, expectations and needs to be addressed. Secondly, the efficiency and effectiveness of the production of these outputs will be an important element of evaluation. Thirdly, the capacity of the system to adapt to new environmental challenges – that is, the flexibility of the system – should be taken into account.

Table 1.4 Mechanisms of a system.

Mechanisms	Purposes	Remarks
Authority and decision-making	To distribute/delegate authority and to formalize how decisions are made and determine objectives.	Especially important as a clear chain of command will avoid unproductive and competitive structures.
Coordination procedures	To ensure that the different parts of a system or a subsystem cooperate in an efficient way.	Present both within subsystems and between subsystems and systems.
Management of resources	To allocate resources in the most adequate way.	Two main consequences can be expected: lower costs and a greater flexibility in the management of resources.
Intelligence process	To collect and analyze significant information for the system and subsystems. To disseminate information to relevant individuals inside and outside the system.	Intelligence, control and communication together form the information system.
Process of control	To develop a mechanism of control to monitor and evaluate actions, services, and outputs.	
Process of communi-cation	To identify and understand needs and expectations of collaborators and external partners. To inform them about projects, actions, problems and results.	
Cultural process	To reinforce the loyalty of actors to the system and their alignment to its objectives.	
Learning process	To guide necessary corrections to systemic processes and actions.	

1.3 A COHERENT SYSTEM OF DECISION-MAKING

The decision-making process is one of the mechanisms central to the operation of a system.

1.3.1 The basic decision-making model

Decisions must be made because problems have to be solved. Hence, the first step when making a decision is to recognize there is a problem in distinguishing symptoms from causes, and to define the problem as precisely as possible in order to develop adequate solutions. The next step consists of listing and evaluating different options and making a choice; the third phase concerns the implementation of the solution chosen. An evaluative procedure should complete the process. This is the basic decision-making model that will be discussed in Chapter 2 in more detail.

Even if very rational decisions are admittedly a myth, the development of, and training in, a rational decision-making method offers many advantages. Such a process helps in harmonizing the way an organization approaches decisions, facilitates the cooperation of different organizational actors on the basis of a common method, accelerates operational preparation and implementation, and reduces the risks of poor decisions.

It is important, at this stage, to clarify the nature of the generic method that will be developed in the remainder of Chapter 1. Though the basic decisional frame illustrated in Figure 1.5 – *analysis, decision, implementation and evaluation* – will remain the same when applied to different situations, the modalities will be defined by the types of problems addressed (administrative, operational, human or technical). The collection of information as well as the resources deployed or tactics used, will also be context-specific.

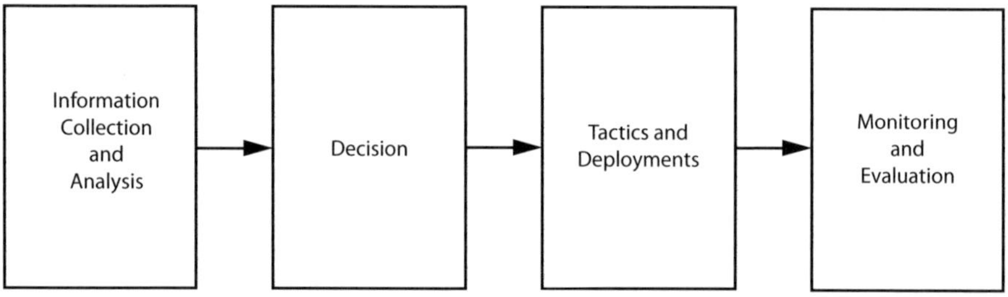

Figure 1.5 Operational decision-making chain.

1.3.2 The importance of coherence

The notion of coherence is of particular importance in the decision-making process. The Figure 1.6 identifies the general conditions contributing to coherence, which derive from the logical interactions between three elements: *the ends, the ways and the means.* The ends are defined here as what has to be achieved in order that a problem may be considered as being solved; the ways are the possible options to achieve the set of objectives; and the means are the resources available. To give an example, a police chief may decide to develop an operation to reduce drug trafficking activity focused around the city train station. The chief sets the objectives of the planned action, defines the tactics to be employed, and allocates the necessary resources to support the operation.

The relationships between ends, ways and means do not assume unidirectionality; it is not always possible to determine first what the ends (objectives) are and then decide on how to reach them before allocating the necessary resources. The ends will have to be defined while accounting for the constraints that may limit the choice of objectives, or the ways to achieve them. Similarly, the choice of ways depends on the types of resources available. However, of key importance is that the solution chosen ensures coherence between means and ways to achieve the stated ends. As in the earlier example, the decision-making process would be incoherent if the chief aimed at completely eradicating drug trafficking in the area without allocating (or acquiring) sufficient resources.

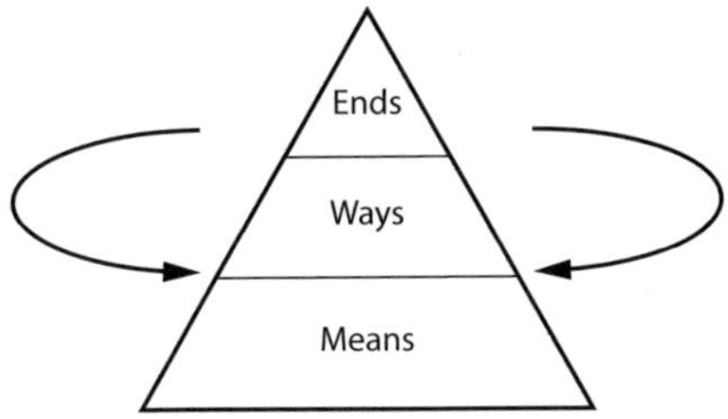

Figure 1.6 Three elements that contribute to coherence.

1.3.3 Elements of the decision-making process

The identification of concerned actors

Figure 1.7 refers to the actors involved in the decision-making process. One must go beyond the simple assumption that a number of individuals supply information to a person, who in turn makes a decision to be implemented by other personnel. The reality is more complex and many outside influences can have an impact on the process. Subsequently, one of the difficulties in the decision-making process is to identify all the actors involved – official and non-official[3] – and to understand their roles and influences in facilitating or hindering the decision-making process, whether positively or negatively.

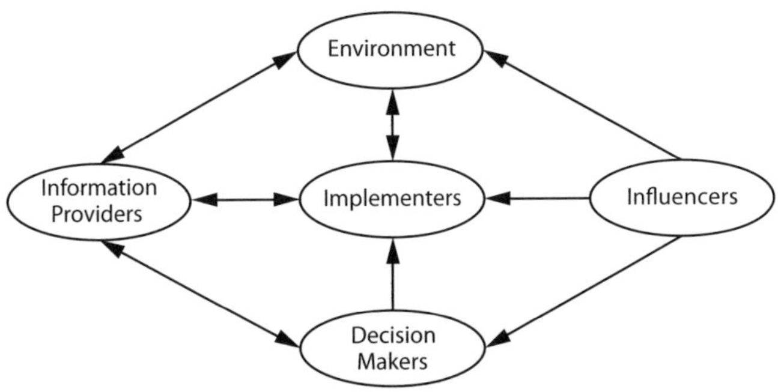

Figure 1.7 Actors involved in the decision-making process.

Starting with a problem or a situation, one can identify the actors who are able to provide information to the decision-maker as well as those who may influence him or her. For example, political authorities can be identified as actors who are entitled to give instructions or who may exert some indirect influence. Likewise, are the actors involved in the implementation of the decided upon measures.

[3] Official here refers to those actors who are recognized or derived from authority within the police. Unofficial refers to those actors in the community or those who may have political or advocacy clout; though they are not an official part of the police, they may have various levels of influence on the decision-making process.

Case Study: A Decision-Making Primer[4]
Good decision-making depends on assigning clear and specific roles in the process. It is thus necessary to identify who is part of the process and in what faculty. The establishment of a function diagram will help that purpose. The five letters RAPID correspond to the five critical decision-making roles: recommend (make proposal), agree (have veto power over the recommendation), perform (responsible for the implementation of the decision), input (be consulted on the decision), and decide (have the final say over the decision).

Types of decisions

a. As a means of simplification, it is possible to group decisions into *three broad categories* (Table 1.8).
b. However, decisions must be analyzed and handled according to *the level at which they are made* (Table 1.9).

Table 1.8 Three broad catagories of descisions.

Category	Description
Political decisions	Decisions that establish political aims and allocate resources, therefore determining the limits of the autonomy of operational leaders.
Managerial decisions	Decisions that deal with organizational questions, project management, financial issues, and matters of human resources.
Operational decisions	Decisions that come under the umbrella of the deployment of resources (human, material, financial).

Table 1.9 Description of decision levels.

Level	Description
Political	Decisions that establish the political objectives to be achieved and the framework within which relevant strategies and operational and tactical measures will be developed.
Strategic	Decisions that take into account the broader environment and concern the ends, involving decision-makers at the highest hierarchical level.
Operational	Decisions that relate to specific elements of a process/operation and must support the overall objectives, involving decision-makers who have been specifically delegated power for this purpose, though to be exercised in line with a broader strategy.
Tactical	Decisions concerning specific measures that are necessary or may influence the implementation of higher-level actions, involving decision-makers with recognized, direct responsibility for the specific area upon which the decision impacts.

[4] Paul Rogers and Marcia Blenko, "Who has the D? How Clear Decision Roles Enhance Organizational Performances", *Harvard Business Review*, 84(1), 2006, pp. 53-61.

c. *The level at which they are implemented*: the distinction between the political, strategic, operational and tactical levels remains useful. The key point here is to clearly define who is in charge.
d. *Who makes the decisions*: it is important to establish a clear system of authority, defining who is entitled to make decisions, and the procedures and instruments of coordination. A chain of command, as well as mechanisms for coordination and control should be developed. It is therefore essential to broach the topic of relations between political and operational bodies as tensions between them, or as confusion about respective functions may produce serious challenges.
e. Their *complexity*: for example, the time needed for tackling a problem, the number of actors involved, or the inherent difficulties of any solution.
f. The *freedom* enjoyed by the decision-maker: this refers to the presence or absence of constraints limiting his or her decision-making autonomy.
g. The *justification* of a decision: this alludes to the causes of an action (Table 1.10).

Table 1.10 Causes of an action.

Causes	Awareness	Trigger for Decision	Time Frame	Philosophy
Evolution of the environment	Political or public sensitivity to specific phenomenon. Risk analysis and detection.	Political demands, opinion pressures, need to eliminate or reduce a risk, need to address a problem.	Future	Proactive strategy
Specific situations	Information about a possible future event (official announcement, open sources, informants).	Mission received (*e.g.*, to prepare the security of a conference). Possible risks created by the event. Judiciary reasons.	Future	Proactive operational measures
Incidents	Incident reported or detected.	Need to act (save, protect, arrest).	Present	Reactive operational measures
Assessments	Political requests, public opinion pressures, operational reasons.	Requests for assessing what occurred and identifying lessons learned.	Past	Learning from past experiences

1.4 THE ROLE OF INFORMATION IN THE DECISION-MAKING PROCESS

1.4.1 The purposes of information

Information is a primary element in the decision-making process; information will help relevant actors to understand a problem and support a decision, as well as the subsequent monitoring and evaluation. Information will thus serve different purposes at different stages in the decision-making process. It is possible to illustrate this variance with Tables 1.11 and 1.12, which highlight not only the importance of the time period – past, present and future – and the levels at which information must be collected and analyzed – strategic, operational and tactical – but also reflect the role of information at different phases in the decision-making process.

1.4.2 The time period

The purposes for which information is used are time-related and will differ on this basis as the following chart shows (Table 1.11).

In operations, information at each level, will have to provide the necessary elements to the decision-maker as follows (Table 1.12):

Table 1.11 Time-related purposes of information.

Time Period	Purpose of Information	Information Needed
Past	To reconstruct/to understand what has happened.	Criminal data for investigation; retrospective analysis.
Present	To assess a situation/to monitor action or a project.	Information needed to set up PESTL and SWOT.[5]
Future	To plan for the future and identify trends.	Scenarios Analysis
Posterior/ after action	To evaluate results.	After action reviews Assessments

Table 1.12 The role of information for decision makers.

Decision-Making	Role of Information
Analysis: Key phrase: to help understand	To support the assessment of issues, highlight trends, correlations and interactions between events. To identify who can supply information and who needs to know.
Decision: to list and evaluate options Key phrase: to help decide	To supply the elements for determining options.
Implementation Key phrase: to monitor	To monitor action. To supply feedback.
Evaluation Key phrase: to learn	To provide data upon which to base an evaluation.
Anticipation Key phrase: to be prepared	To help identify new risks and opportunities.

[5] PESTL is a method of analyzing the environment in looking at its political, economic, social, technological and legal features. SWOT is a way of assessing an organization in identifying the threats and opportunities it faces and its strengths and weaknesses. These tools will be described in more details in the "Second part of the operational decision-making process: the decision" in Chapter 2.

1.4.3 Types of information

It is also important to differentiate between various types of information, such categorization influences the collection of data and subsequent dissemination (Table 1.13).

Table 1.13 Description of information types.

Type of Information	Description
Routine information	Information that an organization must continually collect and analyze in order to track the evolution of its operating environment, or of recurrent phenomena.
Ad hoc/specific information	Information to be collected in order to provide support for a decision in a specific domain. For instance, a commercial company wanting to launch a new product will undertake a market study; a police force facing a large and potentially violent demonstration will need to obtain information on potential troublemakers and targets.
Key information	Information to be urgently transmitted during operations without passing through the usual chain of command, potentially enabling operational chiefs to decide upon and implement immediate measures.
Statistics	Information collected in order to establish the evolution of a specific phenomenon at different periods of time.

1.5 THE ROLE OF INFORMATION IN PROACTIVE AND REACTIVE STYLES OF POLICING

There are limitations to the operational use of information gathered by police. Internal structures may prevent a smooth flow of information within a force. At the same time, the division of police forces within a country, even within a region, hinders, for different reasons, the communication between them. "Linkage blindness" is a general term coined by Egger in 1984[6] that expresses this systemic myopia.

Some new models explicitly recognize that the fluidity of information and the rigor of its treatment are key aspects of policing. These models fully integrate an analytical capacity as a central component. In particular, intelligence-led policing has been adopted by many police forces as a means of achieving a better understanding of security challenges, in order to proactively address them or reduce their impact through an efficient deployment of resources based on sound strategies. This proactive attitude, viewed as a management philosophy, can be considered as a response to the limits of reactive repression, in particular of the investigative process. This has aided in shifting greater attention to other ways of combating crime. The logic of the system consists of a better integration of information into decision-making processes. Crime analysis and criminal intelligence analysis play an important role in an intelligence-led system.

---

6 Steven A. Egger, "A Working Definition of Serial Murder and the Reduction of Linkage Blindness." *Journal of Police Science and Administration*, 12 (3), 1984, pp. 348-355.

Intelligence-led policing (ILP) is considered a philosophy or a business model. Indeed, this approach has important consequences for the whole organization of police forces, the training of police officers, the system of decision-making and on the deployment of resources. It also offers a basis for the evaluation of operations, which can be analyzed in the context of an evolving criminal environment. The National Intelligence System deployed in the United Kingdom constitutes one of the best examples of ILP in action.

Case Study: The NIM[7]

The National Intelligence Model (NIM) represents an effort to promote effective intelligence led-policing on a national basis and to standardize intelligence-related structures, processes and practices across all police services in England and Wales. It is essentially the design for a comprehensive "business process" to rationalize and systematize the ways in which the police service handles information and makes key decisions about the deployment of resources.

Products

The NIM is built around the production and use of four broad categories of product:

- Analytical products;
- Intelligence products;
- Knowledge products;
- System products.

Description of products

Analytical products

The full range of these products are as follows: results analysis, crime pattern analysis, market profile, demographic analysis, criminal business profile, network analysis, risk analysis, target profile analysis, operational intelligence assessment.

Intelligence products

Some of the analytical products listed above are combined or developed into intelligence products, which may be broadly subdivided into strategic assessments, tactical assessments, target profiles, and problem profiles (hotspots or crime series). These are used to inform decisions on the prioritization of problems and targets, to allocate ownership of problems to particular staff or units, and as the basis for planning of operational responses.

---

7 Tim John and Mike Maguire "Rolling Out the National Intelligence Model: Key Challenges", in: Karen Bullock and Nick Tilley (eds), *Crime Reduction and Problem Oriented Policing*, Willan, Portland, 2003, pp. 38-68.

Knowledge products
These are products that will inform future development of the Model itself and maintain quality within it.

System products
These are designed to ensure that appropriate technical and computer equipment is available to the Model for its effective operation.

Tasking and coordinating groups

The key users (and in some cases, producers) of these four types of product are the core bodies which drive the whole process, tasking and coordinating groups.
This process takes place at three defined levels:

- Level 1 Covering local issues (primarily at Basic Command Unit level).
- Level 2 Covering cross-border issues (at force and inter-force level).
- Level 3 Covering serious and organized crime, on a national or international scale.

ILP was preceded by other policing approaches that made more explicit the need for a sound treatment of information. For instance, community-based approaches integrated the idea that "problems" should be systematically identified in a collaborative way with communities. The implementation of methods such as "SARA" (Scanning, Analysis, Response, and Assessment) was seen as a key aspect of this perspective.[8]

Problem-oriented policing is often seen as a policing strategy in itself, which rests on the development of crime analysis. Problems do not stop at "community problems" because the police also identify links within their own information in order to detect security problems that the population may not perceive. For instance, when criminals are mobile, local communities experiencing victimization see only a very small part of the problem, while the police may detect the existence of a series of crimes active on a broad territory.

One of the most famous attempts to implement an entire system centered on the treatment of information is the *Compstat System* developed by the city police of New York, based on a SARA-like methodology. Compstat involves the gathering and analysis of information in a timely fashion and achieves rapid and flexible deployment of resources. The "compstat meeting" puts pressure on decision-makers through an assessment of information, objectives, strategies and resources in an environment susceptible to vigorous debate between participating actors. Global statistics and compstat meetings are tools for maintaining control over a decentralized system.

However, while the role of information has become increasingly recognized within policing environments since the 1990s, the integration of intelligence into

8 See Chapter 2 – Decision.

policing strategies has been slow and sometimes challenging. Reactive and urgent responses and a lack of vision have often combined to deter development of such processes. At the same time, police organizations tend to have distinct experiences in dealing with crisis and responding to isolated incidents where action is at the heart of the process. In proactive policing (anticipating events through crime analysis), there are recognized difficulties in concretely using information for making decisions and operating; in reactive policing, there are noted weaknesses in information processing. By formalizing the operational chain, it is possible to consider these weaknesses in a single framework for decision-making, relevant to both proactive and reactive styles of policing.

1.6 TRAPS IN THE DECISION-MAKING PROCESS

The decision-making process will never be fully rational, it is susceptible to distortion by different factors; some of these factors are within the control of the decision maker while some are not. The issue of individual control will be dealt with in greater detail in Chapter 2.

1.7 SUMMARY AND KEY POINTS

- A system is an ensemble of elements and actors, interacting mutually and with their environment, organized to fulfill certain objectives.
- The perimeter of a system can be drawn according to its actors, identified by their contribution to its finality.
- A systemic approach assists in understanding and accounting of the consequences flowing from the evolution of the systemic environment. New threats and new opportunities may appear and compel the system and its different elements to adapt.
- These changes may impact upon the goal of the system, relevant actors (e.g. the number of missions), their interactions, and systemic mechanisms.
- The decision-making process is one of the most important mechanisms for making rational decisions and for ensuring coordination between different actors.
- This decision-making process can be divided into four clear phases: information collection and analysis, decision, tactics and deployments, and monitoring and evaluation.
- Any decision made must reflect coherence between ends, ways and means.
- There are several types of decisions; differences depend upon the level at which they are made, their complexity, their time period and/or the autonomy of the decision-maker.
- Information is integrated in each phase of the process and supports specific needs: it helps to frame the problem in the analysis phase; it supports decision-making; it facilitates the monitoring of implementation; and it supplies the data necessary for subsequent evaluation.
- At each phase of the decision-making process, bias, prejudices and flawed methodology may detract from the outcome.

Chapter 2

The Operational Chain

Objectives of the chapter

This chapter will further develop the different phases of the operational decision chain as defined in the previous chapter: information collection and analysis; decision; tactics; and evaluation. The chapter will consider the practical implications of each phase and illustrate them through real-life examples.

2.1 INTRODUCTION

Representing the operational decision chain in visual form, one can identify four distinct phases reflecting the horizontal dimension of the process; which refers to the logical order in which a decision is made.

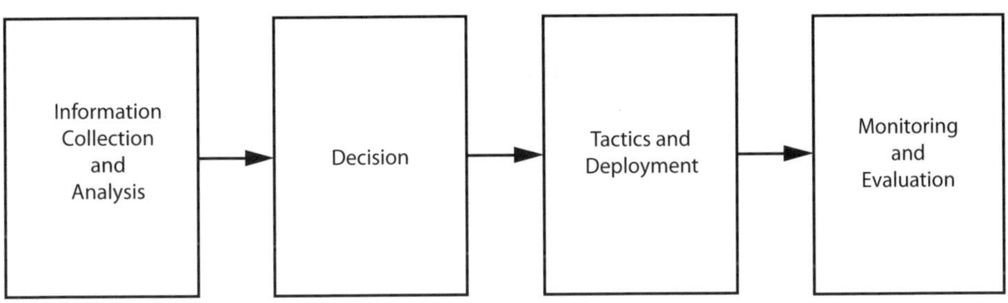

Figure 2.1 Operational Decision-Making Chain.

Each of these phases can be linked to a specific methodology (or methodologies) that represents the vertical dimension of the operational chain (Table 2.2). The main operations employed in the context of the decision process are then presented in Table 2.3.

Table 2.2 Methodologies of the decision chain.

Phase	Methodology
Gathering of information	Intelligence cycle
Decision	Decision model
Deployment of resources	Tactical principles and methods
Evaluation	Audit, assessments

Table 2.3 Operations employed in the decision process.

Phases	Actions
Definition of the problem	Understanding the problem, the identified need, the mission received. Analyzing the problem Formulating the problem
Decision-making	Thorough assessment of the problem as formulated in the phase above. Objectives to achieve and acknowledgement of constraints (*e.g.* resources, legal). Identification and evaluation of possible and authorized actions. Choice of a single option.
Implementation	Plan of action for the execution of the decision. Orders Execution
Monitoring	To track to what extent actions taken are in line with objectives.
Forward planning	To identify possible evolutions of the situation. To prepare reactive measures.
Evaluation	Assessment of the results and the process.

2.1.1 The function of information in each phase

At each phase of the decision-making process, information is necessary to highlight a problem, analyze the risks and opportunities associated with potential options, support a decision, supply the data for assessing potential developments and/or the results of actions taken.

The relations between those who provide information and those who make decisions is a complex one, encompassing the following aspects: how to define the information to be provided, and how to collect, analyze and present it in order that it is useful for the decision-maker. It is also important to consider how efficient structures are established for handling and integrating information into the decision-making process.

The importance of a thorough identification of the actors, the establishment of a clear chain of command and the use of a structured decision process.

There are three main issues to be addressed:

a. Firstly, the *identification* of all those who should be – directly or indirectly – involved in the process, including their roles and responsibilities.
b. Secondly, the *relationships* between the different actors involved in the process (defined by the chain of command).
c. Thirdly, a *structured decision process* involving many actors who may belong to different agencies, and have different roles and responsibilities. Political leaders may play a role, while the whole process may take place over a long period of time. Different types of meetings will need to be organized in order to inform, assess, decide, control or evaluate. The structuring and management of these meetings will determine the efficiency of the process.

In summary, one can observe that measures must be taken in different areas in order to ensure a coherent decision process. The different areas relate to the process itself; the supporting organizational structures (linked to implementation); the identification of the actors, with clarification of their roles and responsibilities; strategic or tactical decisions; methods; and the handling of information. Correspondingly, coherence will only be achieved if the overall management of the process and the principles of the systemic approach described in Chapter 1 are respected, referring to the mutual reinforcement of the ends, ways and means (Figure 2.4).

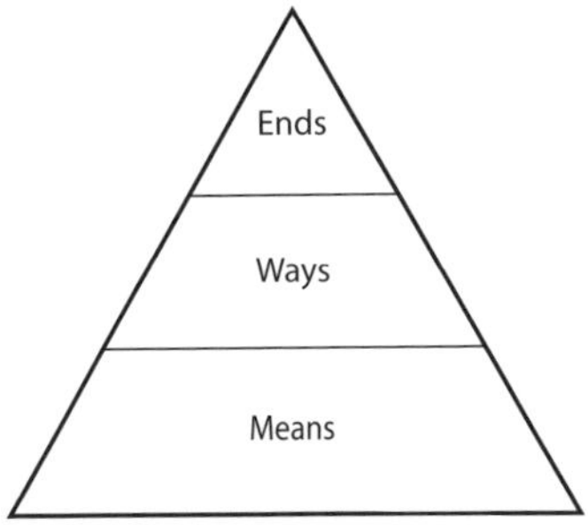

Figure 2.4 Coherency between means, ways and ends.

PART ONE OF THE OPERATIONAL DECISION-MAKING CHAIN: *INFORMATION*

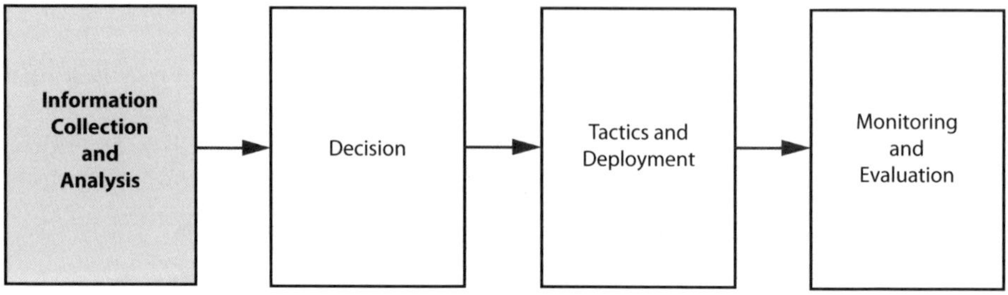

Figure 2.5 Information collection and analysis is the first stage of the information cycle.

Objectives

This Section examines in detail the treatment of information through the "intelligence-cycle" and certain variants of this process. It takes the viewpoint of the intelligence profession that has emerged in police organizations. This will be used to highlight some challenges in establishing a common understanding and credibility between the producers and users of information and intelligence.

2.2 STRUCTURE OF THE SECTION

The operational chain demonstrates how decisions are informed by data gathered and analyzed in a timely fashion, and presented in an appropriate form. When such information is utilized in an operational context, the term "intelligence" is frequently attached. Police organizations have adopted this concept from the military; the analogy emphasizes that pertinent, timely and carefully analyzed information is crucial for the development of a strategy. However, police requirements and needs differ markedly according to the nature of decisions made, the type of information that is collected, the type of problems to be analyzed, and how intelligence tasks are generally distributed across an organization.

The operational chain requires that decisions-makers make *well informed* decisions. *Well informed* can be understood to mean decisions based on the best available knowledge, which results from the structured collection and treatment of accessible data at different tempos, and for different aims. An individual, a unit within an organization, or a multi-agency system covering several jurisdictions or regions may each aim at meeting such standards in order to improve their respective capacities to utilize available information rationally and practically, for the purpose of decision-making. This "organizational" aspect will be discussed in greater depth in Chapter III. The present chapter is focused predominantly on questions of *method*. Notwithstanding this, some references will be made to certain organizational aspects.

A broad consensus exists around the suitability of the "intelligence process" – or "intelligence cycle" when applied at the systemic level – for capturing conceptually the nature of information processing. However, a variety of distinct processes run in parallel in a single system. Some run only for a certain time (*e.g.* the duration of a specific investigation), while others are fixed (*e.g.* via sustained monitoring of risk factors). Each system addresses specific problems (*e.g.* monitoring high volume crime) and assists decisions for various purposes: tactical, operational or strategic. Despite this diversity, the concepts of the intelligence process and intelligence cycle remain valid and analytically useful in most situations. As a result, these concepts represent required knowledge for those intending to manage or analyze information linked to crimes and broader security problems.

Nonetheless it must be acknowledged that the intelligence cycle is an idealized description. Indeed, Ratcliffe notes that he has "yet to meet experienced analysts who work under the strict regime of the cycle."[1] Real-life processes exhibit greater parallel activity and less linear sequencing in the progression through each phase captured by the cycle.[2] Therefore, the cycle should be adapted flexibly to specific and concrete situations. Likewise, it should be noted that some alternative models correspond better to the complex reality of crime analysis where *routine information* is concerned (for analyzing repetitive incidents), or when problems are scrutinized for a sustained period. In all these situations, there is a continuous flow of information that must be absorbed for a particular duration.

Unfortunately, fundamental process issues often remain neglected in a context where information technologies are flourishing and attracting attention. Impressive innovations provide tools for supporting elementary tasks, such as mapping crimes, or the capacity to review vast amounts of data. However, these basic tasks form part of a more global methodology still largely to be described. Frequently, a gap exists that is poorly managed. Systems are often imbalanced at a methodological level by the prevalence of unscientific approaches (that is, based only on beliefs), too narrowly specialized (*e.g.* too exclusively shaped by the sub-fields of psychology, geography or computer science) and characterized by a systematic focus on technology itself (presenting only pins on a map). This important point will be further developed in this section.

Another contemporary fallacy is to overestimate the capacity of information-processing and analysis to respond to the exact wishes or expectations of a decision-maker. Criminal reality is hard to grasp with difficulties of access, as well as deficient information. In hierarchical relationships between a decision-maker and an analyst, there may be temptation to appease authority rather than elaborate hypotheses from structured inferences based on reliable source material. This chapter will also emphasize practical limits and challenges to overcome when developing an organizational intelligence capacity, as well as factors potentially contributing to success.

---

1 Jerry Ratcliffe, *Intelligence-Led Policing*, Willan, Cullompton, UK, 2008, p.105.

2 Peter Gill, *Rounding Up the Usual Suspects: Developments in Contemporary Law Enforcement Intelligence* Ashgate, Aldershot, 2000.

2.3 THE FLOW OF INFORMATION AND INTELLIGENCE CULTURE

This book has continued to reaffirm the crucial nature in information for making good decisions. Consequently, the flow of information within an organization has emerged as an important factor in influencing operational success. However, it is still common to observe problems related to linkage blindness, caused by a combination of human, organizational, methodological, technical and legal aspects. Obvious information-sharing failures have been particularly consequential in certain situations, that attempts have been made to relax previously rigid restrictions on the distribution of information from a traditionally strict "need to know" and "right to know" basis towards a global "need to share" attitude.

At the same time, information overflow has also increased dramatically through the generalized use of electronic devices in routine activities. This evolution has contributed to a new form of linkage blindness: relevant signs are now not easily detectable within huge quantities of data. In a context of information proliferation, privacy also constitutes a significant concern. Thus, in this paradoxical situation between not enough and too much, frequently sensitive information, the "right to know" and "need to know" rules remain valid. They must combine with a "need to share" approach rather than being replaced by it; indeed there is a subtle balance to be found. This brief background illustrates the need for the development and implementation of better models for structured information processing.

2.4 ROLE OF INFORMATION

Information underpins the decision-making process in many different situations and for very different aims (the organizational aspects will be examined further in Chapter 3).

Ideally, in these disparate situations, three broad types of objectives are recognized, corresponding to three different forms of analysis:

- *Strategic*: to support the development of a policy according to a range of security problems over a mid- or long-term period. (*e.g.* strategic analysis may assess illegal immigration or organized crime over a certain period or geographic area).
- *Operational*: to respond and assist in the management of operations and plans (in line with policy derived from a strategic assessment) by deriving specific strategies/operations according to what is observed. (*e.g.* such analysis may point to new immigration movements that necessitates the devising and implementation of a specific strategy and the set-up of an operational task force according to the objectives and means).
- *Tactical*: to meet the need of an immediate response when a specific situation requires such action. (*e.g.* tactical analysis may help determine directions for an investigation).

Chapter 3 describes how specific organizations may separate tactical and strategic functions or the collection and analysis of information as a reflection of their size and competencies, or the geographical area covered. A distinct separation between the different forms of analysis, however, is usually not seen in reality, and there are still many perceptions of what strategic, operational or tactical processes constitute. Categorically, one may consider a continuum between decisions that necessitate immediate response and long term priorities decided by policy makers. In the case of small organizations, some analytical units may well be engaged – in one way or another – in all three forms of analysis.

Case Study: Levels of analysis based on a specific situation
A young man was murdered in a train station. This homicide was perpetrated by two other young men who wanted to remove an electronic device from the victim. The emotion across the region was very high. Media reported inaccurately the idea that trains and railway stations were unsecure, youth delinquency was increasing and the police should immediately respond to this intolerable situation. The analytical units of the police were required to produce an assessment about the evolution of youth delinquency and violence in stations and trains (strategic). This analysis had a significant political impact. As an immediate measure, it resulted in the establishment of a task force where criminal analysts were engaged in a daily follow-up of the current situation in order to guide targeted operations (operational). Finally, specific investigations were launched with analysts supporting investigators' work (tactical).

2.5 INFORMATION PROCESSING

Analysts work on raw material for producing intelligence products, similar to a manufacturer who may need basic material in a raw state for producing an end product. Intelligence is therefore the result of processes that consist of recognizing and collecting relevant data, and transforming the raw data collected into a form suitable for making decisions.[3] This transformative process is labeled the *intelligence process*, and the stages involved are traditionally described by the so-called *intelligence cycle.*

2.5.1 From traces to intelligence

Traces are remnants of a presence or an activity. These are sometimes immediately perceptible, though are more often than not latent, or difficult to recognize. For example, traces may be concealed in a huge quantity of data that will form the raw material such as phone tapes, journals, information from the internet, hear-

---

[3] From the DIKI (Data Information Knowledge Intelligence) model, Jerry Ratcliffe, *Intelligence-Led Policing op. cit.*, p. 94. This variant specifies that knowledge becomes effective intelligence only once it has been transmitted and taken into account for decision-making.

say statements, and bank account movements. Raw material is treated in order to extract relevant facts intended to serve the purpose of analysis.

Subsequently, this information is then analyzed by going beyond the basic facts and generating and testing hypotheses. The intelligence produced must then be disseminated in order to be well understood and useful in the context of specific operational requirements and needs. Figure 2.6 illustrates the chain going from the recognition of signs[4] to the accumulation of effective and practical intelligence for decision-making.

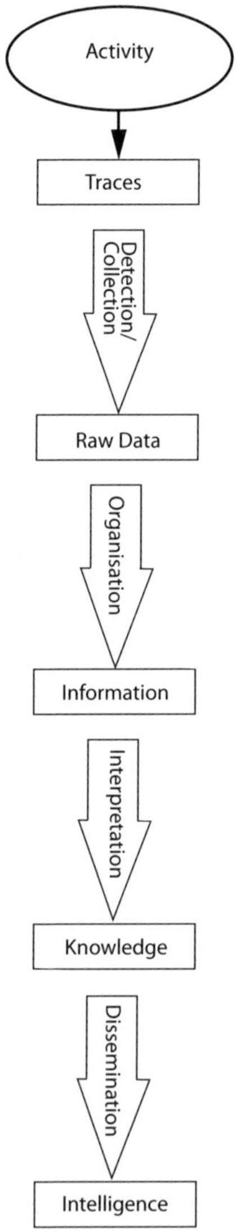

Figure 2.6 Diagram illustrating the chain from traces to intelligence.

4 A trace becomes a sign when we can attribute to it some relevance to the case under investigation.

2.5.2 The intelligence cycle

When integrated into the broader operational chain, these intelligence processes are ideally described by the generic *intelligence cycle*.[5] Within the intelligence cycle, steps are followed in the strict sequence described in Figure 2.7.

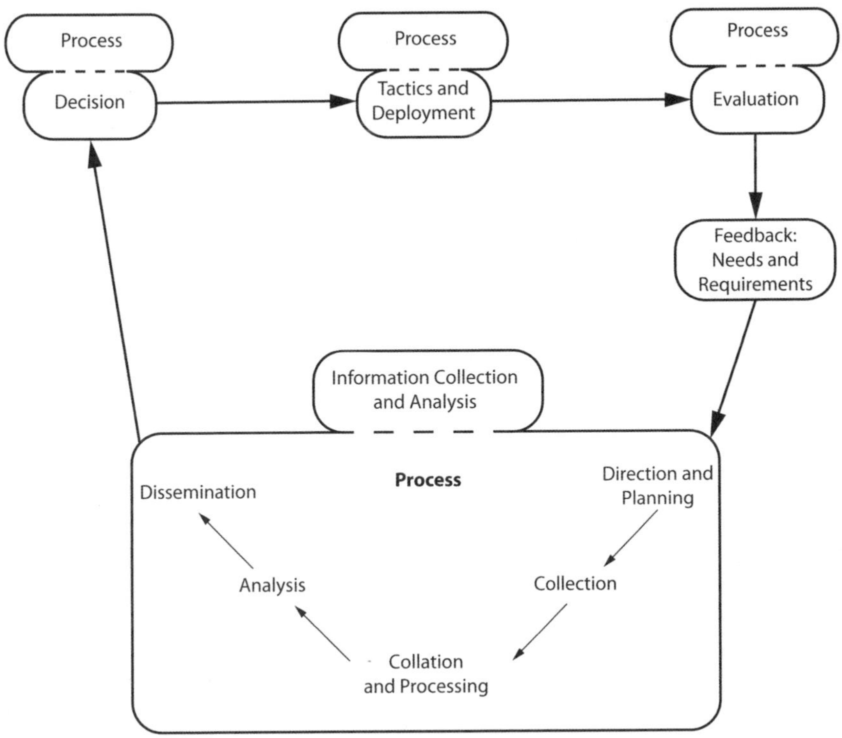

Figure 2.7 Intelligence cycle and operational chain.

Certain key principles underlie this process:

- Maintaining a holistic view of information coming from different sources; the capacity to integrate independent information and connect the dots is a critical aspect of the cycle.
- Independence from other organizational activities, in order to avoid succumbing to group pressure and other traps in decision-making (*e.g.* when an investigative team is pointing to a particular suspect, the analyst may suggest and test alternative hypotheses).

[5] There are several variants of the intelligence cycle, which is expressed here in connection to the operational chain.

- Postponing the analytical part of the process towards the end in order to avoid generating hypotheses too early and thus causing a "tunnel effect" – once facts have been collected, more "inductive" reasoning can begin.
- Understanding that different communities of people may be responsible for specific elements of the cycle. (*e.g.* those who analyze information are not necessarily the same as those who collect data or make decisions – communication tasks therefore assume elevated importance).
- Recognising that new knowledge, or evolution of an existing situation, may require a new cycle to be initiated.
- Avoiding information overflow by focusing on relevant information.

Figure 2.7 illustrates the intelligence cycle and its connection with the operational chain. This representation is an oversimplification, as systems work tacitly with many interlocked cycles. For instance, each phase requires its own information feedback in order to regularly test the coherence between ends, means and ways. A *decision* requires knowledge of what works and what does not, given the alternative choices and information about available resources; *deployment* requires precise knowledge of available resources; the *evaluation* process may also require the establishment of a specific monitoring method and instrument based on the treatment of information. Nonetheless, the intelligence cycle can be of significant assistance in structuring reasoning and guiding the implementation of an organizational-level system (Table 2.8).

Table 2.8 Breakdown of the intelligence cycle.

Task	What	Input	Outcome
Direction and planning	Organizing and planning the process, monitoring its execution according to requirements and needs.	Requirements and needs.	Planning, including collection plan, timeline and resources needed, as well as procedures for the follow-up and monitoring of the execution of the process.
Collection	Applying the collection plan. Evaluating the quality of data collected.	Collection plan.	Raw data with the evaluation of the source and reliability of data.
Collation	Gathering and storing data in a suitable form for interpretation.	Raw data and evaluation.	Database that structures and organises the collected data. Information is ready for examination.

Table 2.8 Breakdown of the intelligence cycle (cont.).

Task	What	Input	Outcome
Analysis	Reasoning by examining and visualizing information. Going beyond the facts in order to infer knowledge that may help in understanding the problem and answering questions.	Prepared data, analytical tools.	New knowledge.
Dissemination	Presenting knowledge in a form that is clear, concise and suitable for decision-making. Communicating products in order to make them translatable into strategies, operations, projects and concrete actions.	Knowledge useful for decision-making (including suggesting actions that may be taken).	Analytical (intelligence – risk analysis) products.
Feedback, new needs and requirements	Reorganizing a new cycle, or adapting the process according to new needs and requirements.	New needs and requirements.	Adaptations of the process.

2.5.3 Implementing the process: further details

Each step of the intelligence cycle is described in greater detail by integrating methodological elements and what one can perceive as factors of success (Tables 2.9, 2.10, 2.11, 2.12 and 2.13 below). This does not assume completeness, as there are many possibilities to develop a system at this level. Only typical aspects that have been personally experienced by the authors or documented have been mentioned. Other elements required for implementation have also been taken into account, such as critiques of computerized tools, required training, and monitoring and assessment. However, these descriptions remain at a rather high level of generality. For instance, crime analysis constitutes in itself a full discipline that is only very generally mentioned here. Indeed, analysis is at the core of the process. Intellectually, it is a very demanding task and necessitates that individuals performing this role conform and adapt to needs and requirements.

Table 2.9 Planning and direction of data collection

Objective	Devise a plan for the collection of raw data as a function of needs and requirements.
Method	• List potentially relevant sources: – Open or closed; – Internal or external; – Government officials or private actors; – Environmental data or related to persons, incidents, as well as signs of other activities; – Possessing a degree of sensitivity (perhaps even protected); – Possessing a degree of accuracy and reliability; – With a definite timeline (must be updated at desired intervals); – Accessible under different formats (electronic, forced choice format, narrative, human); – Requires a cost and different degrees of accessibility; • Evaluate means and timeline for accessing data; • Evaluate data sampling (how representative and complete can the data be); • Specify an adequate level of detail; • Prioritize: information potential, low cost, and easy accessibility are crucial factors; • Continually evaluate the trade off between resources invested in collecting data, analyzing information, elaborating and disseminating intelligence products; • State a clear order and distribute tasks clearly (by using the framework: who, what, when, where, why and how).
Risks	The complexity of the underlying problems, the difficulties in accessing information with available resources and the quality of information potentially accessible, tends to limit the possibility of generating a comprehensive account. Requirements and needs are not expressed clearly enough, leading to the collection of irrelevant data and the overlooking of relevant material. More information does not lead necessarily to better decisions. Collection is not proportionate to the problem at hand and endangers privacy.
Factors of success	Pragmatism, flexibility and the capacity to adapt to new situations. Avoiding blind application of methodology; the size of a project must be adapted to the problem addressed. Emphasis should be on quality of data, rather than quantity. Allowing sufficient time for analysis: often better analysis and good modelling can save resources wasted in collecting useless data.

Table 2.10 Implementation of the collection plan.

Objective	Implement the collection plan.
Method	Follow-up and adaptation of the collection plan, according to progress. Quality of data must be evaluated. Reliability of sources and accuracy of data can be evaluated. There are different methods for this purpose. The use of the '4 × 4' system is very common; it consists of evaluating each piece of information through the following table.

Evaluation of information

Evaluation of the source	1 Accuracy is not in doubt	2 Personal experience by the source	3 Indirect source but corroborated by other information	4 Indirect and not corroborated
A No doubt.	**A1**	**A2**	A3	A4
B Source proved to be reliable in most instances.	**B1**	**B2**	B3	B4
C Source proved to be unreliable in most instances.	C1	C2	C3	C4
X Reliability cannot be assessed.	X1	X2	X3	X4

The 4 × 4 method is often completed by one supplementary dimension, thus becoming then a 4 × 4 × 4 system. The last dimension of this 4 × 4 × 4 system represents an assessment of risks associated with the piece of information (exposure of the source or risks associated to using the information). It is particularly used for exchanging information between law enforcement agencies (*e.g.* international exchanges in Europe). In England and Wales, the National Intelligence Model proposes a 5 × 5 × 5 system which functions on the same basis, but with a finer scale.

Collection of data must comply with laws and rules; should search for the highest quality of data according to the objectives, the management of uncertainties to the greatest extent possible, and the implementation of high-speed and low-cost search procedures.

Recurrent tasks may be identified, formalized and computerized (scanning, automatic recognition of characters, semi-automatic recognition of entities, text mining, etc.).

Table 2.10 Implementation of the collection plan (cont.).

Objective	Implement the collection plan.
Risks	Realization that a plan is unrealistic or the application is hampered by the occurrence of unexpected events. Unexpected flaws in otherwise reliable data sources. The process of assessing information remains highly subjective despite the use of the 4X4 method. Moreover, the analyst is frequently in the situation of not being able to determine these parameters.
Factors of success	High-speed search capability.

Table 2.11 Collating data for analysis.

Objective	Once collected, data must be prepared for analysis; requires structuring, organization and dissemination to authorized analysts.
Method	Develop a model from data collected: • Identify clearly relevant entities, and their relevant attributes. • Identify links between the entities. • Structure the data as a function of the importance of chronological, spatial, or relational dimensions. • Structure the data for potential exchanges of information to maintain comparability. Search for an adequate level of details and balance between free text format and more structured encoding, in order to comply with timeline and ensure that information is suitable for analysis. • Refer to case examples. If a computerized system is available, transform the data into a compatible digital form (scan, encode or import data). Control the quality of the encoding process. Display complex data through mapping, link and timeline, commodity flows, activities and event charts.
Risks	Relevant entities or relations may not be apparent when building the model (this may depend on the collection of new information). Moreover, the problem under scrutiny may evolve, making its representation outdated. In these circumstances, the model may reflect a poorly adapted tool to handle the actual situation.
Factors of success	Not underestimating difficulties related to conceptualization and modelling: this is a high-level intellectual activity that requires careful consideration when building a complementary team. Encoding is an error-prone and repetitive process; motivation and concentration, as well as quality control of the encoding process are crucial. High-speed encoding or automatic importation of data. Keep the model as simple and flexible as possible; knowledge about the problem will evolve and may require a redefinition of the model. A computerized system should be reliable and require as little attention as possible (practical and easy to use).

Table 2.12 Analyzing information.

Objective	Once collated, data has to be interpreted or analyzed in order to suggest potential actions to decisions-makers. Based on this analysis, risk analysis products will be devised.
Method	Results of analysis must be timely and useful, consistent with needs and requirements. Analysis takes several forms: • Strategic, operational or tactical. • Target profiles, criminal activity, profiles, hot spots/routes, recurrent modus operandi, specific events, links between entities, and the search for patterns potentially indicating illicit activities. Analysis is about developing reasoning based on uncertain and incomplete information: • Going beyond the facts, developing hypotheses. For instance, in forecasting the most likely possibility and potential paths of change. • In particular, moving from the effects (what is observed) to the possible causes (adductive and hypothetico-deductive reasoning). • Necessitates generalization from singular observations (profiles). • Main dimensions are link analysis (analyzing links between entities) and spatial/temporal analysis. • May involve detecting patterns in a huge quantity of cross-referenced data (data mining technologies may support this process). • Requires the adoption of critical thinking. A possible method is to frame analysis by the 5W + H methodology: Who, What, When, Where, Why, How (and with what)?
Risks	Due to the uncertainty pertaining to data sources and inference processes, hypotheses may be not valid. Thus, there is a subsequent risk that decisions are made based on erroneous evidence. The time available for in-depth analysis may be insufficient.
Factors of success	Analysis rests on the availability and quality of information and of the use of specific knowledge. Harmonized methodology of work is needed in order to achieve common standards; yet one must be cautious of the concept of *standardization.* Determinism is not alluded to, only a global framework at an adequate level of generality. Consciousness of cognitive biases and an application of critical thinking is required (a scientific attitude in the broad sense). Imagination and adequate knowledge are crucial qualities for generating hypotheses. Special importance should be attached to the selection, vetting and training of analysts.

Table 2.13 Dissemination of knowledge.

Objective	On the basis of the hypotheses developed, knowledge is disseminated with the intent of facilitating decision-making.
Method	Dissemination can be oral or written. Many structures for reporting exist: • Identify the recipients of products and tailor the product accordingly. What are their requirements? Keep in mind the principles of "need to know" and "right to know" relative to potential recipients' functions and responsibilities. • Products should make clear the separation between what are facts versus hypotheses (*e.g.* a map representing crime events and the interpretation of the spatial distribution). • Products may be structured according to Who, What, When, Where, Why and How (5W + H): – What is the problem? – Who is concerned? – How and why: what is the mechanism behind the problem? – With what: what are the instruments and means? – When and where the problem occurs. • Use visualization tools (*e.g.* crime mapping, entity-relation diagrams, and flow charts). • Retain a clear description of the method of analysis in the event that important decisions are made when levels of uncertainty are of primary importance. • Suggest possible lines of actions that are adequate according to the result of the analysis (depending on the organization).
Risks	The transfer of information can also involve unintended transformation; this problem increases in a multilingual environment such as in Europe.[6] False hypotheses lead to poor decisions. Beyond the impacts of the specific measure following a decision, this may lead to a net loss of credibility that is advantageous of analysts. Frequently, problems remain unclear despite analytical efforts. Intelligence products must correspond to what can reasonably be developed on the basis of available information, irrespective of the high demand of precision by decision-makers. Caution should be made against pressures from decision makers to produce hypotheses more precise than possible based on the available information.

6 "Translation is the language of Europe" (famous saying attributed to Umberto Eco).

Table 2.13 Dissemination of knowledge (cont.).

Objective	On the basis of the hypotheses developed, knowledge is disseminated with the intent of facilitating decision-making.
Factors of success	Intelligence is developed for practical use; how intelligence may be transformed to support concrete measures is a major concern. The acceptable level of risk of poor decision-making is evaluated as a function of uncertain hypotheses. Highly-sophisticated intelligence products are avoided in preference for more directly informative material. Abide by KISS (Keep It Simple and Stupid). The KISS principle states that simplicity should be a key goal in design, and that unnecessary complexity should be avoided. The duration of presentations and length of reports must fit with resources available to decisions makers. Ensure that analysts observe their organizational role and focus on the problem at hand.

Table 2.14 Assessment of intelligence measures.

Objective	To evaluate the impact of measures implemented on the basis of intelligence.
Method	When analyzing a reoccurring problem, one should compare the specific situation *before* and *after* action has been taken: • Choose indicators according to the objectives of the program. • What was the situation before implementation of the program? • What is the situation after implementation of the program? • If possible, a comparison should be made with a scenario where no action was taken (a control). • Observe the differences; is it only possible to explain any divergence in terms of actions taken, or do other factors come into play? • Adapt the process if necessary. When the problem consists of a single event (*e.g.* a demonstration) or an investigation supported by intelligence, "lessons learned" should be made explicit via a formal process.
Risks	Selected indicators can conceal the complexity of a problem.
Factors of success	Keep the assessment realistic, but ensure implementation occurs in the context of an understood system that lower-level officers cannot mistake for hierarchical surveillance.

2.6 SUSTAINED ANALYSIS

Many situations require fixed or sustained analysis. High volume crime is regularly scanned in order to detect patterns. Detecting and managing crime hot spots would fall under this category of activity. When a crisis occurs, or criminal intelligence analysis is required in the framework of an important case, sustained gathering of new data and interpretation of the new flow is critical for guiding the investigation.

Case Study: Continuous analysis in the course of a specific investigation

In an important case where a group of 15 persons were suspected of theft, robbery, fraud and other crimes, the continuous processing of new information acquired through interviews or other investigative operations was particularly important in structuring case knowledge, and eventually evidence. Almost 1,000 interviews were conducted during this investigation; a substantial part of them concerned the suspects. Individuals changed their explanation according to the state of the investigation and information available to investigators. This is a typical attitude – suspects tend to find alternative explanations that are favorable to them when a previous explanation is no longer coherent within the context of newly collected evidence. The management of the different versions of the same protagonists therefore became important. When interviewing a person, the investigator must be made aware of all preceding versions of the answers of the person in question, in order to evaluate and anticipate how the interview should be conducted. This was made possible thanks to the devise of a continuous process that lasted during the whole investigation (about 5 years) and included the implementation of a specific database. The work of the analysts was eventually crucial in assisting investigators in structuring evidence and preparing documents and visualizations for the court process.

When intelligence processes are established in order to absorb a continuous flow of information, the ideal intelligence cycle does not adequately represent the parallel treatment of the different phases. Figure 2.15 depicts such a process. Though featuring the same steps as the intelligence cycle, it emphasizes the need for the continuous update of available knowledge in order to obtain an accurate picture of the current situation as new information becomes available. Although this is an operational process, the knowledge gained can be used both at a tactical level (general knowledge for supporting specific investigations) or at a strategic level (global knowledge on some crime mechanisms).

The improved model serves as a good illustration of intelligence processes devised for continuous analysis of information. For instance, the systematic scanning of high volume crime, violent crimes, repetitive fires, and processes related to surveillance and risk analysis as well as a specific investigation (see case study above) fit this interpretation. Moreover, the model introduces an important new concept: the memory. The memory is an abstract description of what is known by the organization about the problem of interest at a certain time. It should be noted that the model neither mentions at this stage, computerized systems nor the way of organizing the process. This is an abstract description that must be developed into an appropriate organizational structure, equipped with computerized systems that can support reasoning within this framework.

2.6.1 Collection

The collection of new data is shaped by knowledge about the current situation. For example, which patterns have been identified? What is the profile of the crimes? How have they developed? Who are the suspected perpetrators? What are the needs and requirements for managing the crisis, including from an investigatory perspective? For instance, when a burglary is reported, a patrol intervenes and collects basic information. The scene may be processed by crime-scene examiners developing forensic case data and a description of the *modus operandi* of the offender. Other sources of information may come from individual informants or witnesses of various types.

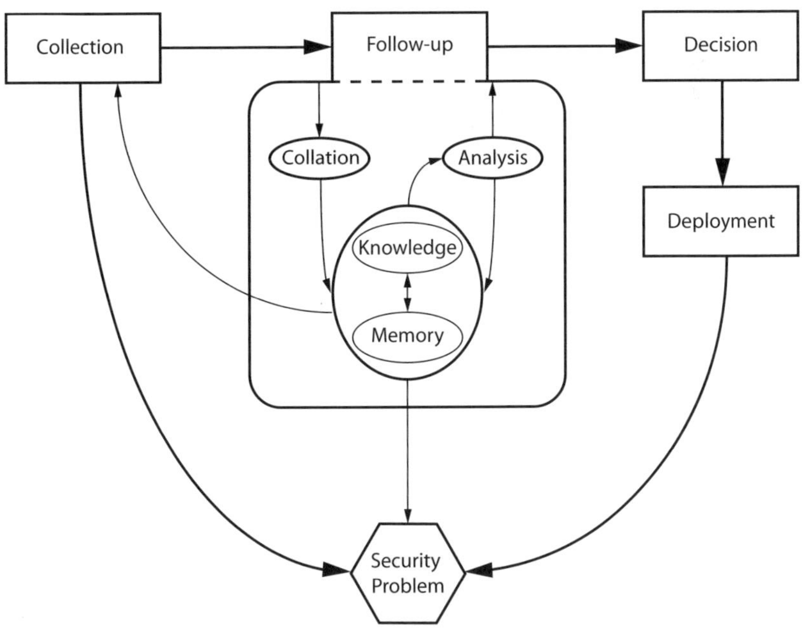

Figure 2.15 The process of information absorption and continuous analysis. Relevant signs or traces are collected and integrated into a memory representing already existing knowledge concerning a security problem. This knowledge is continuously updated and interpreted in order to influence decisions and adapt strategies as a function of the evolving situation.

2.6.2 Collation

When a crime-analysis structure exists, this new information is interpreted, potentially identified as relevant and connected to existing information already recorded (*e.g.* where a link with another burglary or series of burglaries is detected or assumed), and finally integrated into a structured memory. The logical structure of the memory is crucial. It may be organized around concepts such as *repetitive crimes*, the *link* between incidents, or more generally, the regrouping of information into *crime problems*. The definition of these concepts involves some fundamental problems: how does one represent links between incidents, when a degree of uncertainty always remains about the accuracy of such connections? How does one update the memory, when new information contradicts previously assumed statements? These are all new questions of modeling still to be investigated, which are not adequately answered by contemporary research and experience.

2.6.3 Analysis

In parallel, the structured collection of data (the *memory*) is continuously analyzed in order to detect new patterns of criminal activities, analyze relevant profiles, test hypotheses (*e.g.* are two similar series connected?) and ensure coherency. Regularly, such analysis results in knowledge to be disseminated in the form of intelligence products. This leads to a discussion about whether the situation in question should be addressed at the strategic, operational or tactical level, as understood through analysis.

There are many applicable forms of analysis, ranging from the detection of hot-spots, to the analysis of tendencies, or the determination of a profile for a specific reoccurrence of criminal activities (a general description useful for forecasting subsequent events).

2.7 DISSEMINATION

Intelligence may be disseminated in various forms (Table 2.16). The chosen form in any particular case must be suitable for decision-makers; meaning that the intelligence product must be prepared in order to target the appropriate management level:

1. Who are the intended recipients?
2. What was the question asked (informational requirements and needs expressed)?
3. What product will meet those needs?
4. How and when will the intelligence be disseminated?

2.7.1 Examples of possible intelligence products

Intelligence products can be structured in many different ways, usually tailored according to specific recipient needs and organizational styles. The following examples represent some of the most common possibilities.

Table 2.16 Various forms of intelligence dissemination.

Levels	What	Possible Frequency
Example at a strategic level		
Strategic risk analysis	General overview, forecast of the future.	Annually.
Periodic reports	Specific type of security problem	Upon request.
Examples at an operational level		
Operational risk analysis reports	General document that guarantees that reliable knowledge of the situation and risks is developed. The document provides assistance to plan, control and monitor at a regional level. It may contain different visualizations, such as hot spot geographical information.	According to the evolution of the situation and the type of security problem; every few months.
Periodic reports	Assessment report of criminal activity.	Four times a year.
Establishment of a list of known and pervasive phenomena and target profiles	Description of profiles of activities and offenders.	To be generated once, and then updated with new information.
Examples at a tactical level		
Tactical assessment reports	An update assessing identified problems across a region.	Weekly.
List of possible criminal activities and relevant evidence	When a suspect is arrested, rapid information is necessary to develop a rough outline of their potential activity. This is necessary for interviewing purposes or to support a decision to keep the individual in prison.	Upon request.
	Once the profile of the potential activity of the suspect is understood, pointing to relevant evidence.	According to the determination of potential criminal activity.
State of an investigation	Visualizations (*e.g.* charts, diagrams) Criminal analysis reports.	Occasional.
Specific incident	Key incident information (*e.g.* the interception of a suspect's vehicle).	Immediate dissemination to all relevant recipients when available.

Strategic reports

In various situations, strategic reports may be required by a police organization. Very often, when a salient incident attracts public attention, an immediate demand emerges for police to provide a general picture of where the incident took place.

For instance, in the event of a bizarre homicide, the public would immediately demand to know the entire picture for similar events. Unfortunately, while this should not constitute the main use of strategic analysis, relevant resources are very often employed in such a reactive way. In contrast, in a well-structured intelligence-led system, strategic reports are regularly required to support policy development and thus retain a pro-active focus. While there are many alternative possibilities, such reports may be structured according to the diagram of Figure 2.17.

STRATEGIC REPORT

1. **Foreword**
   1.1. Introduction
   1.2. Objectives
   1.3. Executive Summary
2. **Method**
   2.1. List of sources
   2.2. Information and analysis process
3. **General situation**
   3.1. International situation
   3.2. Situation in neighbouring jurisdictions
   3.3. Internal situation
4. **Description of threats and vulnerabilities**
   4.1. Identified threats and vulnerabilities
   4.2. Possible developments
5. **Conclusions**
6. **General recommendations, proposals for priorities and possible countermeasures**

Figure 2.17 One possible structure for a strategic report.

Problem and target profiles (strategic/operational)

Problem and target profiles are synthetic descriptions that represent a set of criminal activities or disorders, subject to routine analysis; for example, mechanisms behind organized forms of criminality, groups of perpetrators, or profiles of persons repeatedly victimized.

The intended audience of problem and target profiles will often be the membership of the organization at large, as these products represent general knowledge useful at every managerial level. Thus, such documents form part of a knowledge repository that may be maintained in a data bank and may be referenced during various training programs or recalled when a police organization decides to prioritize the targets in question. Such analysis may also occur on an occasional basis, when there is a need for greater depth in the understanding of specific criminal mechanisms or groups of perpetrators. Table 2.18 shows an example of a form that can be used to collect, collate and update these problems and target profiles. Here, it is used for describing the activity of a group of burglars.

Table 2.18 A form used to update problems and target profiles.

Profile	
What	*Very general description of criminal activities:* Hundreds of burglaries in apartments organized by a specific organization employing young people.
Who	*General profile of the offenders:* Mostly young females.
When	*Period of activity:* During the day, from Monday to Friday.
Where	*Scope of the activity:* Apartments in upper floors, in city-centers.
How/ with what	*Modus operandi:* With a screwdriver, breaking the door of the apartment, selecting jewelry in the bathroom and the bedroom and removing bank notes.
Remark	The offenders generally circulate with specially-plated used cars. The organization exploits child offenders in order to escape severe punishment.

Weekly operational reports

The crime analysis units of police organizations scan the criminal environment on a daily basis and making a weekly report. Operations dealing with already known problems are also monitored and specific incidents may require special attention. These analysis typically feed a weekly report where positioning of patrols, specific surveillance activity, or investigative leads are decided. A possible structure for a report (intelligence product) supporting such a meeting is shown in Figure 2.19. For each point, the product may take the concrete form as defined in Table 2.20.

WEEKLY OPERATIONAL MEETING REPORT

1. **For each known problem and phenomenon**
   - 1.1 Evolution, new incidents.
   - 1.2 Assessment of the situation, possible developments.
   - 1.3 Devise and monitor operations.
2. **New problem detected**
   - 2.1 Description.
   - 2.2 Possible developments.
   - 2.3 Recommendations.
3. **Specific incident**
   - 3.1 Description.
   - 3.2 Consequences.
   - 3.3 Recommendations.

Figure 2.19 Possible structure for the weekly report.

Table 2.20 Model for elaboration of points in reports.

Observation	Interpretation	Recommendations
Burglaries at the border Burglaries are perpetrated along the border; the following localities have been visited by the perpetrators (...).	The perpetrators leave their cars at the border in country A, and access the properties in country B by foot. The perpetrators generally operate during the night of Monday to Tuesday.	Organize surveillance along the border along the border next Monday night, if possible in cooperation with border guards of country A.

Key information (tactical)

When key information is known, there is an urgent need to disseminate it to the right people. Within an organization, it is important that such information is presented in a consistent and intelligible form in order to alert officers to its importance and urgency. A practice implemented within a police force that has been labeled *INFO-FLASH*, is a name easily recognized by most protagonists (Figure 2.21).

INFO-FLASH

1. Phenomenon in reference.
2. Recent developments.
3. Proposition for immediate measures to be taken.

Info-flash		
Transport of stolen goods in vehicles prepared for exportation.	Stolen goods have been found hidden between the seats and the driver's door.	Control relevant vehicles more intensively.

Figure 2.21 The structure and an example of an info-flash.

Case Study: Key information in a specific case of credit-card theft

A suspect car and the first numbers of its license plate were observed by a witness just after the theft of a wallet containing a credit card from a shopper at a shopping centre parking lot. This information was immediately conveyed to the crime-analysis unit, which recognized the activity of a known group of thieves. The unit rapidly distributed an Info-Flash directly to all available patrols, recommending the patrols focus their efforts on shopping centers of the same region. After a very short time, one patrol arrested the perpetrators at another shopping center as they prepared to commit another theft.

2.7.2 Visualization

An image is worth a thousand words, and visualization helps communicate a message. There are several typical ways of visualizing crimes and security problems. Remembering the 5W + H model *(Who (with whom), What, When, Where, Why and How (with what)),* it is not surprising that most forms of visualization are generally a composition of maps (where), relational diagrams (entities and relationships – who, what, with whom, with what), as well as chronologies or temporal representations (when).

Types of visualizations

a. Maps

Crime mapping is a traditional way of communicating crime data and is recognized as a very simple and relevant method of representing security problems for decision-makers, since most of the time, both criminal activities as well as deployment of police resources are largely constrained by space and time. There are many types of crime maps – such as pin maps, chloropleth maps, thematic maps, or 3-D maps[7] – with their own advantages or disadvantages.

A very useful type of map to assist decision-makers is a hot-spot map (figure 2.22). It is generally recognized that criminality tends to concentrate in time and space to form "hot spots." Studies have demonstrated that patrols can have a direct effect on crime reduction by focusing their effort on these areas. Hot spot analysis has therefore become a frequently-used technique. The figure below shows an example of crime hotspots calculated within a small city over a certain period of time.

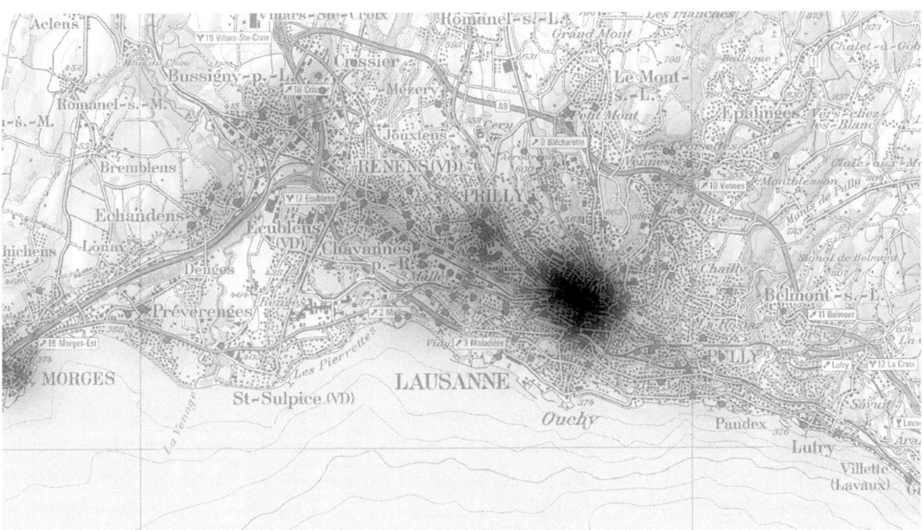

Figure 2.22 Example of a map of hot spots within a city.

7 See for instance Rachel Boba "Crime Analysis with Crime Mapping", 2nd Edition, Sage, London, 2009.

b. Relational diagrams

In all criminal cases, relevant objects and individuals interact in the context of a series of incidents. Representing the relations between these "entities" is often very pertinent, especially when the number of entities and relations reach a level of complexity that exceeds the capacity of a normal human brain to manage (Figure 2.23).

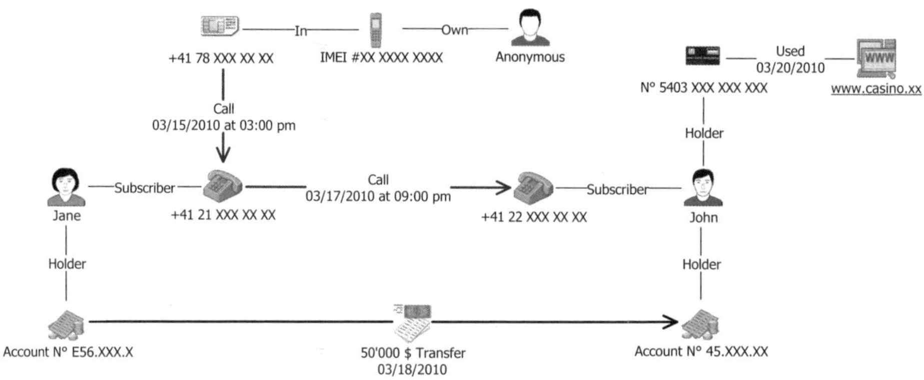

Figure 2.23 Example of relational diagram.

c. Temporal representations

Flow diagrams have become a traditional method for representing all kinds of transactions: a sequence of telephone calls, movement on bank accounts, or selling of goods may be illustrated in this way.

d. Combinations

Sometimes, more than one dimension may be relevant and should appear on a single map. Figure 2.24 is an example of a spatiotemporal representation of incidents. The time line at the top of the screen and histograms give an indication of when the incidents occurred, while the map provides the spatial dimension. Surprisingly, despite the evident usefulness of such tools, these kinds of visualizations are not well integrated into computerized geographical information systems, nor designed to be used by non-specialists. Indeed, the example below (Figure 2.24) comes from an ad-hoc computerized system developed specifically to cross-analyse mobile phone or GPS data from a spatiotemporal perspective.[8]

Another possibility is to combine relational and temporal analysis – typically useful when a series of incidents needs to be visualized. The example depicted in Figure 2.25 shows how this kind of combined approach was used to represent a series of repetitive murders assumed to be perpetrated by the same offender.

---

8 The figure has been extracted from the 'visualist' software developed by Quentin Rossy. The software is used in police forces, especially for the systematic treatment of mobile phone data. Quentin Rossy, Visualizing Crime Data, PhD Thesis, University of Lausanne, under evaluation.

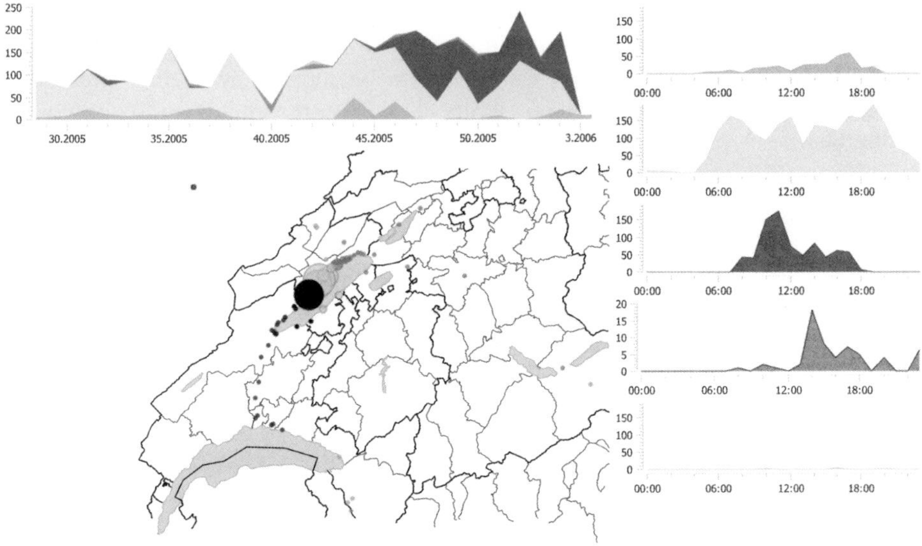

Figure 2.24 Example of a spatiotemporal representation.

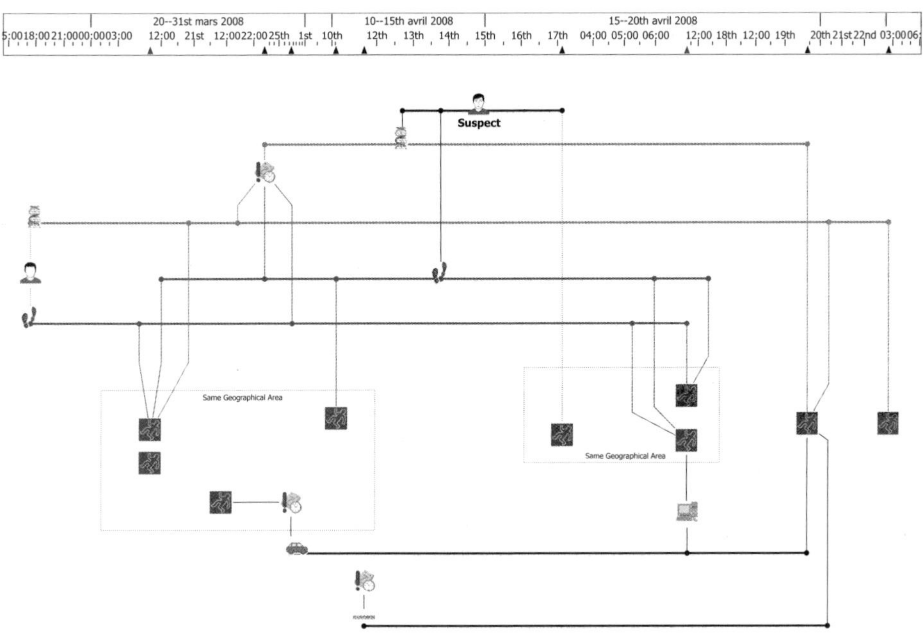

Figure 2.25 Relational and temporal analysis.

Poor visualization and representational biases

Behind nice diagrams, many potential biases can intervene to result in representational caricatures of problems, rather than a balanced view. Tufte[9] provides a remarkable account of these biases, while also recommending simple approaches to enable the development of efficient visualizations.

Case Study: Dissemination of the results of a survey

The chart (Figure 2.26) shows the results of a survey. It has been published in a report produced for politicians. The study intended to measure the degree of satisfaction of foreigners (born inside or outside the country) with the performance of police.

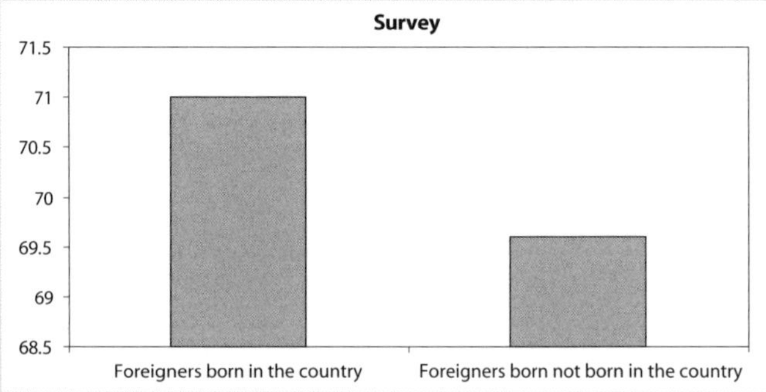

Figure 2.26 Survey results.

At first glance, this chart suggests that foreigners born in the country are more satisfied about the police than foreigners not born in the country. There are many political consequences to this interpretation. However, there is no statistical significance to the observed difference between the results for both categories. The bias appears because of the origin of the vertical axis, which commences at 69.6% and gives the illusion of a marked gap in satisfaction. The histogram is an unlikely attempt to deceive the reader, as such diagrams reflect the default representations obtained when using traditional software. The size of the sample (N = ?) should also have been mentioned on the histogram.

It is very easy to find glaring examples of distorted visual data in newspapers and even in scientific journals. When representing data and information, one should bear in mind a seemingly trivial, but often ignored statement: one can represent only the information available. What is unknown cannot be represented. While crime analysts are generally well aware of this bias, once information is disseminated to non-specialists and decision-makers, problems may arise.

---

9 Edward R. Tufte, *The Visual Display of Quantitative Information*, Graphics Press, Cheshire, Connecticut, 2001.

Case Study: A potentially misleading visualization

In the example depicted in Figure 2.27, the individual at the center of the diagram was under surveillance (in circle). As a result, an abundance of information on the suspect was collected (visible in the connections mapped), irrespective of its significance. Yet, other persons ultimately proved more crucial to the outcome of the investigation.

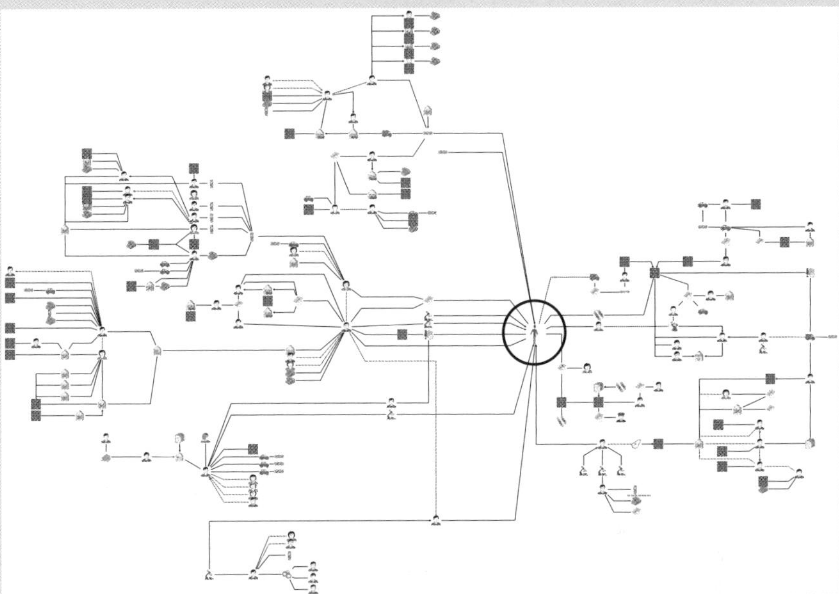

Figure 2.27 Visualization can be misleading.

Another problem arises once information or intelligence has been disseminated: the broader context that surrounds a visualization may be lost. Visualizations are intended to support analysis of a problem (where the user is the analyst), versus the use of visualizations to explain or disseminate information to other partners, these functions should not be mixed.

Case Study: Losing control over interpretation

In the context of community policing, crime analysts were asked to produce maps for use at public meetings, representing crime problems across a specific locality. In one case, a meeting was held shortly after a very mobile serial burglar had struck at least 15 times during one night in a single village. Hence, the map displayed a spectacular rise in burglaries in the village in question. However, though this rise was very easy to explain, those presenting the maps were not crime analysts, and hence were not able to discuss nor put into context the situation described on the map to local authorities. This contributed to exaggerated fear and reactions across the local population.

2.7.3 Oral presentations

In certain circumstances, oral presentations may be preferred to written reports. Table 2.28 gives some recommendations for when this form of communication is used.

Table 2.28 Recommendations for oral presentations.

	Checklist	Key aspects
Preparation	1. Do the analysis. 2. Prepare the presentation. • Identify to whom it is intended. • Know where it will take place and organize the room. • Determine the technical resources available. • Know the precise time available. • Write the presentation. 3. Learn the presentation.	A perfect understanding of the problem is obviously a precondition. The presentation should be formal and the room must be organized accordingly. The comfort of those attending may be an important aspect (e.g. chairs or heating in the room). The duration of the presentation should be well-controlled and practiced in advance. Learning the presentation beforehand will make the delivery more fluid and enable the speaker greater freedom from the text.
Structure of the presentation	1. Short introduction (welcome). 2. Objective of the presentation. 3. The method. 4. Results obtained. 5. Interpretations. 6. Discussion. 7. Conclusion.	The presentation begins by trying to attract the attention of the audience. The objective should be clearly stated, and the audience also clearly identified in order to target the message to be disseminated. Concentrate on a limited number of messages to convey. The audience should always know which point the presenter has reached in the presentation.
Material supporting the presentation	1. Always visible, minimal text, do not overload with tables. Use key words. 2. Avoid a series of bullet points in favor of visualizations. 3. Ensure coherence throughout (not changing colors for representing similar aspects).	Unreadable texts and tables will detract from the intelligibility of the presentation. The equipment used must not hide the speaker.

Table 2.28 Recommendations for oral presentations (cont.).

	Checklist	Key aspects
Broad approach	1. Leave time for questions and discussion. 2. Leave time for the public to read the slides and interpret graphs. 3. Do not speak too rapidly. 4. Have a "plan B". 5. The presentation should be targeted to the entire audience, not to a particular individual. 6. Do not mention the weaknesses of your own presentation. 7. Do not go beyond the time allotted.	Never try to answer a question you have not understood or for which you do not have an answer (the speaker cannot know everything). While the speaker knows his or her presentation, the audience will be exposed to the logic for the first time – for instance, time is needed to understand graphs. The objective of the presentation is not to demonstrate the skill or knowledge of the analyst; rather, it is to convey a message. Remain flexible in case of unexpected events (technical, timing). Do not concentrate your attention on a particular person (*e.g.* on the chief or the screen); the presentation should be targeted to the entire audience.

2.8 THE INFORMATION TECHNOLOGY FALLACY

Information technologies have flourished over the past decade and the potential applications for criminal intelligence are numerous. At the same time, while such technological advances have resulted in some successes, there have also been disappointments in view of the level of resources invested. Computer systems dedicated to criminal intelligence have displayed many weaknesses – partly due to fundamental difficulties in conceptualizing requirements, needs, and underlying methodological models. In criminal intelligence, models attempt to capture a largely unknown, fluctuating and informal kind of activity with uncertain data. Moreover, the policing environment is one of constant pressures and is not a suitable place for designing conceptual approaches to problem analysis. Most knowledge is tacit and acquired through *experience*. Thus, a clear methodological gap exists and has left the space for many ineffective applications of new technology.

The complexity behind the development of computerized systems has often been underestimated, which has induced certain fallacies.[10] Though building a computer system requires a holistic and interdisciplinary approach, some managerial attitudes serve to simplify the problem. This is illustrated when an intelligence capacity is built around information technologies or specific software, rather than

[10] Gary T. Marx and Ronald Corbett have identified 9 forms of what they call "techno fallacies": "Critique: No Soul In The New Machine: Techno Fallacies in the Electronic Monitoring Movement", *Justice Quarterly*, 8, N. 3 (1991), pp. 399-414.

grounded in a sound methodology and tested body of knowledge. The fallacy pervades multiple organizations and is largely fed by a frenetic business environment created around information and communication technologies. This state of affairs is particularly evident in the field of criminal intelligence. In several cases, sophisticated and impressive computer systems have been presented directly to police managers; once purchased, the staff is required intensive (and costly) training. For some managers, the problem of developing criminal analysis capacities is then considered solved.

Such underestimation of the difficulties involved is particularly evident when considering the prior background necessary for adequately interpreting spatiotemporal information. This body of knowledge has been built from years of research in criminology, with studies dedicated to questions such as why and how criminals travel – by following certain patterns or concentrations of crime – and how to discern these patterns when analyzing crimes. Therefore, a diverse range of theoretically-informed inferential approaches must form the basis on which computer software is developed. Buying a database is of no use if individuals have a limited understanding of the models that underpin its operational use. Computer software is inseparable from the inferential activities it is intended to support.

Unfortunately, the managerial mistake of confusing the tool and the method is recurrent. Many criminal intelligence structures have been developed around the expensive acquisition of computerized tools, and not around sound methodological approaches. Indeed, such tools may also unnecessarily limit reasoning by concentrating the attention of users on available software functions. More often than not, this situation bestows upon the analyst/operator the status of "specialist": that is, "the one who knows how to make the computerized system work." Of course, such a perception rests on false rationality, as the ability to operate software says nothing about one's aptitude for analyzing criminal intelligence.

Another strategy for developing software dedicated to criminal analysis is to employ an iterative approach. The principle rests on the rapid development of a prototype, which will then evolve in the workplace when confronted with real case situations. So-called "agile software development techniques" structure this approach within a methodology that appears particularly suitable for dealing with ad hoc and evolving problems. Indeed, in many organizations, analysts themselves develop their own tools. Usually, such software is not encouraged by management on account of a perceived loss of control over the development process.

Unfortunately, in these circumstances managerial attitudes tend to coalesce around two different extremes:

- Considering the developed software as the result of "tinkering" and buying impressive, but expensive and unusable software from the market.
- Considering the computerized system as the reason for success and trying to sell it.

Both approaches will probably contribute to the death of the computerized system, since the software is inseparable from the environment in which it is deployed.

Case Study: Development of criminal analysis software

In one instance, database and analytical tools were developed iteratively within the workplace over a period of more than ten years. As a result, the computer software was perfectly integrated as an extension of the work of the analysts in question. The software was used every day, and was adapted according to fluctuations in the crime pattern itself. While the success of the software was self-evident, it was due largely to the methodology used, the team of analysts that systematically requested for modifications, and productive cooperation with a partner university in maintaining and developing the model behind the software. Nonetheless, because of its success the creators planned to sell the system to other organizations, regardless of the fact that the software worked well because it was tailored to the specific functioning of the team and formed part of a non-transposable organizational culture. The personnel and the methodology were the essence of the success, not the tools supporting the approach.

Table 2.29 summarizes the arguments and outlines possible approaches for making the most of technological tools.

Table 2.29 Making most of technological tools.

Objective	To utilize computerized systems in support of information processing and the implementation of risk analysis.
Method	Computerized systems must be integrated into the whole system: the organization, personnel, processes and existing instruments. Instruments must support all steps of the process, from collecting data to analysis. Data mining technologies may support analysis where applied to large quantities of data. Computerized systems must be secure, flexible, user-friendly and adaptable, according to the evolution of criminal activities, needs, requirements and technology.
Risks	Expensive and inadequate computerized systems monopolize attention. Though ambition is worthwhile, it may disguise complexity; for instance, aiming for full interoperability or avoiding redundancy in a database may hide chains of difficulties. Long development cycles may lead to systems that are already outdated upon completion. Software, computers and related techniques wrongly become the center of the attention. Development and integration times are incompatible with needs, and the evolution of technology and the criminal environment. Each computer system must permanently and progressively be adapted to the rapid evolution of an environment (must be considered as a perpetual task).

Table 2.29 Making most of technological tools (cont.).

Objective	To utilize computerized systems in support of information processing and the implementation of risk analysis.
Factors of success	A whole risk analysis must be devised before choosing adequate computerized systems and not the other way round. Computer systems are tools to support a methodology to an end. The real potential of the technology is to be evaluated, particularly when envisaging data mining or other ambitious techniques. Building computerized systems require an intensive modeling activity, not just a patchwork of individual experiences. Human and computer contributions to the global resolution of problems have to be adequately balanced.

2.9 INTELLIGENCE CANNOT ALWAYS MAKE THE BOSS (DECISION-MAKER) HAPPY

A temptation in producing intelligence material is to make the boss happy with products that go far beyond what information and available knowledge can reveal. Decision-makers occasionally make unrealistic demands; they may also make known in advance what results they would like to receive from the analysis undertaken. Such conditions have the potential to cause the worse attitude on the part of analysts: to bring to the "customer" what they expect to receive and not what can be extracted rationally from the original information. The relationship between analyst and decision-maker is thus an important consideration in the dissemination phase of the intelligence process.

Case Study: A Murder in the train station

The example of the murder of a young man that occurred in a train station has already been presented. The emotion caused by this event was very strong, generating significant pressure on police to make trains and stations safe after these places had been negatively portrayed by the media. In order to respond to public expectations, a group of ten policemen were directed to identify, arrest and bring to justice individuals who where disrupting the peaceful environment of train users. Analysts were supposed to inform these policemen when and where they would find such offenders. Data was difficult to gather because most events were not reported. However, beyond the very high cost of gathering data, it was also soon clear that no real pattern could be discerned. Given the rarity of relevant events in relation to the space and number of trains to observe, it was clearly not possible to gather the precise information expected. Nonetheless, there was a lot of pressure on the analysts, who were criticized for their inability to produce good intelligence. The analysts were asked to "refine" their analysis and provide "valuable" evidence to the operational officers.

Beyond the nature and accessibility of data, analysis of the problem, and the precision of evidence and solutions that can be offered, analysts may also be faced with the pressure of unreasonable demands. Occasionally, questions asked (framed as the requirements for decision-making) might constitute excellent PhD subjects to be completed by a researcher in three to five years. Such unrealistic expectations can introduce unwarranted biases into the intelligence process.

Decision makers are under great pressure in a police system because the operational reality imposes many choices at a rapid pace. Decision makers generally feel uncomfortable when the treatment of a known crime problem must be postponed until resources are again available. Thus, it is possible for decision makers to prefer not knowing, rather than being in the situation of knowing and feeling the obligation of acting, but not being able to do so. The next case study is an example of tensions resulting from this attitude, and how it may be dealt with.

Case Study: Relationships between the chief of investigations and an analyst

A crime analysis unit had been developed in a police force. Tensions between investigators and this unit were frequent. One day, the chief of investigations, and other experienced investigators, had a discussion with an analyst. "You are useless," said the chief of investigations, "you only give us more work by detecting new crime problems, and my team already cannot handle the essentials." The response of the analyst was the following: "We only provide you with information and intelligence. From this, the choices are yours and we will never criticize them, even if you decide not to follow our suggestions. But it is not a good solution to close our eyes towards a real problem for reasons of lack of resources." From this point, a radical change of attitude of members of both units was noticed. It reduced tensions and developed real synergies between investigation and crime analysis.

These are only examples of how the relationships between producer and user of intelligence are important in the decision-making process, and how crime analysis and intelligence unit are difficult to implement.

2.10 EVIDENCE-BASED PRACTICE AND INTELLIGENCE

Evidence-based practice is developing at a rapid pace in many fields, from medical disciplines to some criminological and police activities. This approach seeks to shape practices of intervention on the basis of previous high-quality evaluation studies about what works and what does not. For instance, in which conditions are crackdown[11] operations efficient? Would video surveillance be effective in a

[11] Crackdowns (CD) are repressive and dissuasive operations where the police concentrate substantial resources, for a limited time, on targets (areas, persons) clearly identified as causing serious problems (see section Chapter 2 – Tactics and Deployment of Resources).

particular context? There is no doubt that for police, such use of "lessons learned" from previous operational experience, or application of the research insights evident across a vast and diverse criminological literature may help to introduce more knowledge into the decision-making process (see also the discussion in Chapter 2 on: Evaluation – what works, what does not work). At present, however, this knowledge is still not effectively integrated into day-to-day practices. This necessitates a shift in attitude from police – in developing greater links with academics – and for criminologists to develop new communication skills in order to present results in a form comprehensible to decisions-makers; that is, beyond their own communities in peer-reviewed scientific journals.

A vast debate over methodological aspects continues: what distinguishes a valid from an invalid study? Other dangers exist when an exaggerated attitude is adopted. For example, when statistical data treated under strict norms are not available for the problem at hand, no decision may be made. Beyond such debates, there is an urgent need to focus on the nature of decisions to be made in a police environment. This will hopefully help to provide a framework where academics and police would be able to discuss the value of an evidence-based approach without concentrating only on the formal aspects, but also on practical requirements; that is the specific needs of the law enforcement world.

2.11 SUMMARY AND KEY POINTS

- Information processing can be generally described through the intelligence cycle.
- There are many intelligence processes simultaneously active within an organization, at different levels and concerning various forms of decisions.
- There are alternative representations of the intelligence cycle in order to take into account the absorption of a continuous flow of information.
- Implementing an intelligence capacity must not be driven by the technology; technology must serve intelligence.
- The role of intelligence is not to make the boss happy.
- Evidence-based practice encourages the utilization of high-quality evaluation studies that may introduce solid knowledge into the decision-making process.

PART TWO OF THE OPERATIONAL DECISION-MAKING PROCESS: *THE DECISION*

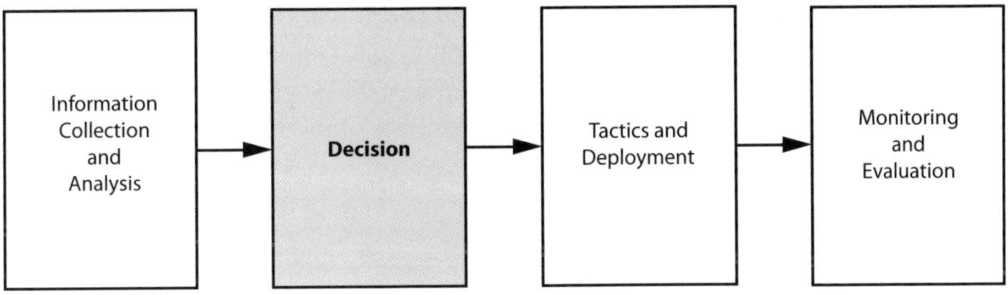

Figure 2.30 Decision phase of the process is highlighted

Objectives and structure

This Section will examine in detail the decision-making process, defined as the logical analytical process leading to decision in the context of operations and/or projects. The methodology can be used in all phases of the decision-making process – including analyzing, deciding, implementing and evaluating.

The chapter begins by presenting the basic rational model of decision-making, before going on to examine more specific models. Subsequently, the chapter describes some potential traps in the different phases of the process and suggests certain tools that can be employed. Finally, the chapter concludes with a summary of key points and factors of success.

2.12 THE RATIONAL MODEL

Theory states that decision-making is a rational process; reality is different. Decisions are never completely rational because decision-makers have neither all the necessary information nor the time and the capacity to identify and evaluate all available options. Decision-makers are also never completely free of personal prejudices and biases or from outside influences.

Although the rational decision-making model remains an ideal-type model, it offers a valuable framework for decision, and supplies a useful method for tackling problems in a variety of situations. Indeed, the method is useful at different levels: it is valuable for structuring the decision process and for organizing the work of a group; it can also be applied to presentations, oral or written, of the results for preparation. Beyond the preparation phase, the same process can be applied in the implementation and evaluation phases of an operation or project. In essence, the model describes how, after a mission has been received or a problem or need perceived or recognized, different options are identified and evaluated in order to make a choice, how orders are given for implementing the solution, and how subsequent action is monitored and evaluated.

It is appropriate to re-emphasize what was stated in Chapter 1: the basic decision model will remain the same in a variety of situations. However, the manner of application will be adapted to specific situations.

Table 2.31 surveys the different operations forming part of the process and methods that can be used to perform them:

Table 2.31 Operations involved in the decision-making process.

Decision-making steps	Problems to solve	Some methods[12]
Understanding of the problem.	To define the problem, the sub-problems, the actors. To identify threats and opportunities, strengths and weaknesses.	• Brainstorming; • Actor mapping; • PESTL; • SWOT.
First measures.	To identify and execute measures that must be taken immediately in order to prepare the next steps, prevent risks and facilitate the process.	• Check-list of the possible categories of measures; • Brainstorming.
Situation assessment.	To assess the factors that may influence the decision to be made and to identify possible options.	• Analysis grid; • Brainstorming.
Decision.	To decide on one option and define the aim.	• Evaluation grid.
Implementation.	To develop orders: who does what, when, with which resources. Execution.	• Operation order.
Supervision of activities.	To monitor the course of action.	• System of control and feedback.
Contingency planning.	To identify and evaluate the potential risks and opportunities.	• Assessments; • Scenarios.
Post-action review.	To evaluate the action/operation.	• Debriefing procedures

2.12.1 Definition of the problem

A solution can only be found if one becomes aware of a problem, a need, or a threat, and fully understands its nature. To adequately and clearly formulate a problem is without a doubt the most important step in the decision-making process; an efficient solution can only be developed if it addresses the root causes of the problem and not only its symptoms – think here of an iceberg, for which at first glance one only observes the tip and not the most dangerous part. Sufficient time has to be allocated for this analysis; whenever possible a participative approach is

12 Some of these methods are described Chapter 2 – Decisions – Tools.

recommended. Some of the important points to be taken into account at the stage of problem-definition (or analysis of the mission) are:

- What is it all about? What is the problem?
- What is the complexity of the problem? Complexity must be assessed as it influences tactics. A simple problem may be solved by a single measure, or a few measures taken by an individual; a complex problem with potentially far-reaching consequences could involve many stakeholders and require complex management.
- Must a decision be made? In some cases, one can also decide not to act. (*e.g.*, a company having established a list of the security risks can decide not to address certain risks because they are covered by insurance).
- In which context does the problem have to be tackled, how does it relate to other on-going projects, problems and operations?
- What is the expected outcome: maximal and minimal?
- What is the necessary timeframe: must a decision be made urgently or not?
- Can the problem be broken down into sub-problems?
- Who are the parties involved (directly or indirectly)?
- What are the opportunities and the risks?
- What are the resources available and do they meet requirements?
- What are the constraints on the decision-maker?

The result of this analysis is a written formulation of the problem, synthesized in a few sentences, which defines the key issues and objectives, and highlights the potential risks and priorities.

2.12.2 The decision

A clearly stated problem assists in developing an appropriate solution. The issues highlighted in the conducted analysis during the definition of the problem are completed in that phase in order to collect all useful information for making the decision. The nature of the defined problem must be carefully assessed, as this may have consequences for the search for solutions. Problems can be categorized in three different broad groups:

- *Generic*: A generic solution applicable to the same type of recurrent problems should be developed.
- *Exceptional*[13] *and generic*: This pertains to exceptions for the organization in question, but not for others, hence consideration should be made towards maintaining communication with other organizations, to observe how they managed similar problems.
- *Exceptional*: In the case where new developments emerge that cannot be addressed within scopes of reference, a specific solution must be established.[14]

---

13 Meaning, different from the standard or norm.

14 For more information about these groups (generic, exceptional and broad, and exceptional), refer to Peter Drucker, "La Prise de Décision". *Harvard Business Review*, Editions d'Organisation, Paris, 2002.

Case Study: Geneva International Conference for Peace in the Middle East
Security was a major issue for the International Conference for Peace in the Middle East, held in Geneva in 1983. The Chief of Police of the neighboring canton of Vaud sent an observation team with a mission to study the security system and obtain all possible information. The documentation collected, and lessons learned, had a great impact on facilitating the preparation and successful conduct of the force under his responsibility during the *Peace Conference for Lebanon*, which was held unexpectedly in Lausanne (a city located 70 km from Geneva) one year later.

In addition, three elements will determine the bounds of the decision and must be clearly stated at the beginning of this phase:

- The *objectives* to be achieved in order that the problem at hand could be considered as solved;
- The constraints to account for (*e.g.* with regards to resources available, legislation or delays);
- The risks at stake; and
- The stakeholders.

The list of possible solutions must be established within this framework.

One should not elaborate more than four options. To identify them, the brainstorming method described in the "tools" section of this Chapter is very useful. The following framework may be used to present the selected options:

- Describe each option in a few sentences.
- State the expected result.
- (Potentially) develop specific criteria: that is, necessary resources, costs, and timing.
- Define the advantages and disadvantages of the option.

2.12.3 Evaluation of options and a subsequent choice

A decision is a choice between different solutions, which all have advantages and disadvantages. Important elements in the success or the failure of the decision depend not only on the appropriate choice, the commitment of leaders, the timing and quality of implementation, but also on other factors. For instance, factors such as feasibility and/or acceptability must also be considered during the options stage. A solution that would endanger the life of individuals would not be chosen when other possibilities exist, even if they involved many disadvantages. Different methods exist to evaluate options prior to a decision, such as:

- Comparisons with the listed objectives and constraints
- Brainstorming
- Ranking through priorities
- Weighting
- A matrix

2.12.4 Testing of the chosen solution

When a decision has been made, it should be "tested"; that is, considered in the context of certain questions or scenarios in order to gauge potential outcomes. Examples of such questions are: does the solution solve the problem and its underlying causes? Is implementation feasible given the resources available? Are the risks involved manageable? What are the possible collateral effects? Whenever possible, the solution should be tested against a scenario or a simulation to evaluate in advance how well it addresses different threats and/or problems that could occur.

2.12.5 Proposal

The solution chosen may need to be submitted to a higher level for decision. Thus, it is of utmost importance to present it in a professional manner. The presentation ought to indicate:

- The solution chosen (and the options not selected).
- Why it has been chosen.
- Resources needed.
- Relevant time-frame.
- Roles and responsibilities (specific to the organization in question).
- Control and reporting systems.
- A communication plan (if necessary).

This basic model can serve as a framework for both oral and written presentations.

The Chapter so far has focused on the generic decision-making model; further discussion will shift to the varied application of this model in different contexts. As will become evident, the basic framework will always remain the same and can be applied in police, military, or business environments. The modalities of application will, however, need to take into account the specificity of individual situations.

2.13 THE MILITARY DECISION-MAKING MODEL

It is important to take into account the military decision-making model, as the police has frequently adopted it (sometimes in adapting it) for its specific operations (planning of the security of large events, for instance). Though there may be slight differences between the military as to how they structure the decision-making process, the basic model will remain applicable. The model can be synthesized in five steps and with two crosscutting activities:

- Five steps:
  - Understanding the problem;
  - Evaluating the situation;
  - Making a decision;

- Planning for implementation of the decision (and if necessary forward planning);
- Executing orders.
- Two crosscutting activities (undertaken throughout the decision-making process):
  - Time planning;
  - Initial measures.

In its specific decision-making process, the military will emphasize certain aspects of the different steps. Usually, the military will spend much time understanding and analyzing the mission if the situation allows. One can generally separate this phase into three stages: the first involves stating the problem, the context of the mission and the time frame in which action is required. The second involves achieving a broad view of the objectives of the task and dividing them into subtasks. The third consists of defining the organization and priorities. The result of mission analysis will be a clear description of the tasks to be undertaken, the objectives to be reached, the sub-problems to be executed (including the degree of urgency) and a general formulation of the problem (see for example the NATO case study featured in Chapter 5).

The estimate of the situation also requires the full attention of the military, often involving thorough analysis of five groups of factors: the mission, time frame, context (environment, terrain), possible actions of the adversary and own courses of action. Key consequences are usually drawn from this estimate, which will highlight the necessary elements to be taken into account in the development of options: for example, where the principal effort must be made. Another important feature of the military approach is the importance allocated to forward planning. General Molkte, Chief of the Great Imperial General Staff of the German Army in the wars against Denmark, Austria and France in the 19th century, noted that planning became obsolete at the first encounter with the enemy. This thinking explains why a military chief will develop a certain number of scenarios in order to retain as much control as possible over unexpected events. Thus, staff will spend time conceiving of potential scenarios in advance, and of the different measures to be taken in each case should such events occur.

2.14 THE POLICE DECISION-MAKING MODEL

Though the police decision-making model is very similar to that of the military, the operating environment, the types of situations potentially faced, and the conditions surrounding execution of decisions differ substantially. In order to focus on police forces in general, one must first try to look at the kinds of decisions that are made at different levels of police organizations. One could categorize these as see in Table 2.32:

Table 2.32 Different types of police decisions.

Categories	Types of decisions	Characteristics	Decision-makers	Methods
Operational decisions	Traffic and incident management.	Generally, an immediate response is required. The number of police officers involved remains low.	Patrol officers. Operational supervisors.	According to the procedures developed.
	Large-scale events.	Complex cases, handled with many personnel.	District/regional divisional commanders.	Structured decision-making process.
	Crisis.	Complex cases involving different agencies, consequences at different levels and with certain durations.	Chiefs Commissioners. Senior executives. Political authorities will also generally be involved.	Structured decision-making process. Coordination.
	Planning for large events.	Importance and complexity of the event. Cooperation of many agencies. Political involvement. Large interest from the media. Time for planning.	Specific staff brought together especially for the event.	Decision-making process.
Investigations	Conduct of the investigation.	Depending on the type of offence.	Prosecutors, investigators and hierarchy of the criminal police.	Structured investigative methods.
Maintaining order	Tactics. Rules of engagement.	Military style operations Highly sensitive because of political, public opinion and media impacts.	Operation chief (with a clearly defined mission and delegation of authority)	Structured operational decision-making process
Prevention	Solutions to be developed in partnership with external partners.	Solutions aiming at preventing, reducing, or eliminating recurring problems of concern to the community and the police.	Police and community.	Participative management and problem analysis techniques (problem-oriented policing).

Table 2.32 Different types of police decisions (cont.).

Categories	Types of decisions	Characteristics	Decision-makers	Methods
Managerial decisions	Definition of the strategies. New project developments. Budgetary issues. Reform processes.	Human factors highly sensitive. Long term process.	Political level and Police Chief or delegate.	Structured decision-making process. Project management.
Technical decisions	Acquisition of new equipment, vehicles, computers, etc.	Costs incurred and technological complexity may be very high. Altering the basic decision once made may be very difficult (important consequences).	Political level and Police Chief. According to the authority delegated within the budget procedure.	Clear definition of the needs and project management methodology.

The complex nature of police decision-making can be best illustrated by considering *four specific applications of the general police decision-making model*: the first in the area of investigation, the second when faced with a challenge to the maintenance of public order, the third when dealing with management issues, and the fourth when engaging with problems of a technical character.

Beginning with the scientific method and its application to *criminal investigations*[15], one can observe (as outlined in Table 2.33) similarities with the rational decision model.

Table 2.33 Similarities between the scientific method and criminal investigations.

Scientific method	Criminal investigation	Rational decision model
State the problem.	What crime was committed?	Define the problem.
Form the hypothesis.	Hypothesize as to possible suspect based on information gathered from witnesses, physical evidence gathered at the crime scene and motive (if determined).	

15 James W. Osterburg, Richard H. Ward, *Criminal Investigation, a Method for Reconstructing the Past*, Anderson Publishing, Cincinnati, 1992, p. 36.

Table 2.33 Similarities between the scientific method and criminal investigations (cont.).

Scientific method	Criminal investigation	Rational decision model
Collect data by observation and experimentation.	Seek out pertinent records. Continue the effort to locate and interview additional witnesses. If possible, obtain from each potential suspect's person (home or automobile) samples, for comparison with physical evidence discovered at the crime scene. Interview witnesses again if necessary.	Evaluate the situation.
Interpret the data as a test of the hypothesis.	Review and evaluate the evidence available relative to making a case – for and against – for respective involvement of each suspect.	Identify options.
If the data supports the hypothesis continue to collect additional data (as a logical consequence of the hypothesis).	Seek additional evidence (possibly through follow up measures) that test hypothesis of scenarios.	Examine the options.
Draw conclusions and elicit a theory if sufficient indicative data is available.	If evidence amounting to probable cause has been developed, arrest suspect. Continue to structure evidence.	Choose one option.

The second variation on the police model relates to the *operational decision-making process* for managing a large and potentially violent demonstration (Table 2.34).

Table 2.34 The operational decision-making process.

Steps	Purposes	Examples
Problem definition (analysis of the mission)	*Analysis of the mission* i What is it all about? ii Expected outcome. iii. Freedom of action. iv. Resources (own and possible reinforcements).	To develop measures to assure the security of a large event organized to take place in two weeks, which planners intend to be peaceful but could be troubled by some violent elements.
	Consequences • To identify what has to be done and when. • To establish a general vision of the problem and to articulate constitutive elements.	Problems relating to traffic, public order, investigatory activities and legal frameworks will have to be addressed.
	• To define the attribution of powers and tasks.	The uniformed police shall deal with... The criminal police shall deal with ... The ... shall deal with ...

Table 2.34 The operational decision-making process (cont.).

<table>
<tr><th>Steps</th><th>Purposes</th><th>Examples</th></tr>
<tr><td>First measures</td><td>To increase the decision-makers freedom to launch immediate and specific measures that could save time and facilitate future actions.</td><td>To define the information requested.
To make recognizance of the area of operation.
To alert, to mobilize (staff, specialists).
To convene a meeting (assessment).
To inform (those who are concerned).
To identify contacts to be made.</td></tr>
<tr><td>Timing and organization</td><td>Tasks and their schedule.
Chain and structures of command.
Establishment of specific working groups if necessary.</td><td><table>
<tr><th>When</th><th>What</th><th>Who</th></tr>
<tr><td>12.3/1400</td><td>March</td><td>Organizers</td></tr>
<tr><td>12.3/1200</td><td>Roads closed to traffic</td><td>Chief of traffic</td></tr>
<tr><td>12.3/1000</td><td>Security in place</td><td>Chief of operation</td></tr>
<tr><td>...</td><td>...</td><td>...</td></tr>
</table></td></tr>
<tr><td>Evaluate the situation</td><td>Analysis of seven factors with potential consequences for the operation:
• Mission
• Legal basis (important for police decisions)
• Area of action (terrain)
• Weather
• The potential actions of offenders
• The potential actions of police (a list of different options)
• Time constraints</td><td><table>
<tr><th>Other actors</th><th>Police possibilities</th></tr>
<tr><td>peaceful demonstrators</td><td>to facilitate and direct their march.</td></tr>
<tr><td>violent troublemakers within the march</td><td>to collect information;
to apprehend them before the beginning of the march;
to apprehend offenders during the march;
to cooperate with the organizers' security service.</td></tr>
<tr><td>violence outside the march</td><td>to organize surveillance and observation;
to secure areas at risk;
to constitute a mobile force for rapid interventions</td></tr>
</table></td></tr>
<tr><td></td><td>Synthesis: consequences</td><td>Consequences:
• To organize the collection of information
• To establish contact with the organizers to define cooperation before and during the march
• To constitute a mobile reserve force</td></tr>
<tr><td>Decide</td><td>To choose between options.</td><td>To choose, from the possibilities outlined – taking into account available resources – the actions that would best address the most dangerous potential action of other actors.</td></tr>
</table>

Table 2.34 The operational decision-making process (cont.).

<table>
<tr><th>Steps</th><th>Purposes</th><th>Examples</th></tr>
<tr><td>Give orders</td><td>To indicate the course of action for the mission.</td><td>Structure of an operations order
a. Situation
• Activity
• Subjects
• Associates
• Offenders: possibilities and risks
• Mission
• Resources
b. Intention
• Summarizes the general mission plan
c. Missions
• Orders given to subordinates
d. Specific instructions
• Behaviour
• Weapons and ammunition
• Use of weapons
• Clothing and equipment
• Special equipment
• Transportation
• Medical support
• Meals, breaks, reliefs
• Communications-radio, telephone
e. Locations
• Chief of operation
• Sub-unit leaders</td></tr>
<tr><td>Control</td><td>To monitor the action and to take corrective measures if necessary.</td><td></td></tr>
<tr><td>Prepare</td><td>To consider how events could potentially evolve in order to take devise preventative measures.</td><td>Scenario planning
Preparation of if -then planning
<table><tr><th>Code</th><th>Hypothesis</th><th>Decision</th><th>Time</th></tr><tr><td>ADO</td><td>Violent incidents in the shopping area</td><td>To send the mobile force</td><td>10'</td></tr></table></td></tr>
</table>

Figure 2.35 illustrates the process followed in assessing a problem.

The third example deals with decision-making in relation to management issues.

As the broad framework remains the same as the first two examples illustrated, it is more useful to focus here on the question of problem definition, which is different in nature (though still serves as an interesting point of comparison to other cases). This is developed in Table 2.36.

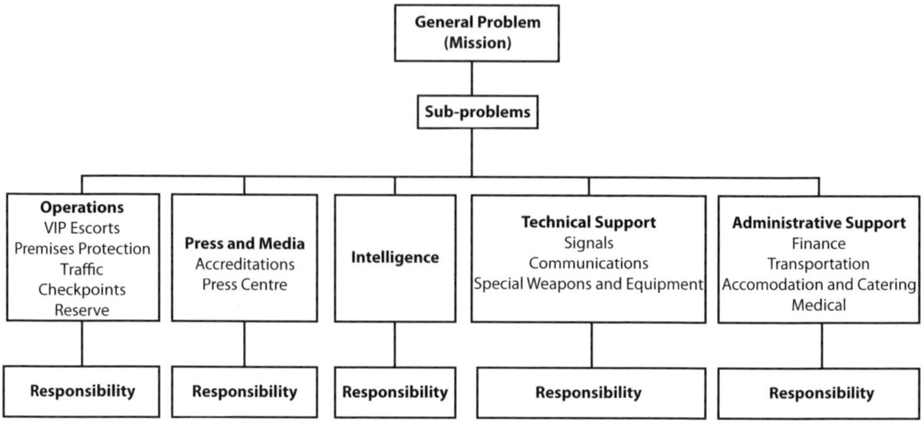

Figure 2.35 Assessing a problem.

Table 2.36 Definition of the problem.

	Steps	Developments	Remarks
1	Situation	Definition of the problem.	The causes of the problem should be described (not be confused with symptoms).
2	Time	Degree of urgency, available time, schedules.	
3	Links with other projects	Identification of potentially related projects or issues that should be taken into account.	It is necessary to highlight these links in order to eliminate redundancies and develop possible synergies.
4	Lessons learned in other projects	Identification of what might have led to success or failure in other similar cases	One tries to avoid repeating errors and to capitalize on elements of success (in assessing if the conditions are the same).
5	Identification of persons/ groups concerned	Who is concerned? Who can provide information? Who should be informed? What kind of information to provide/to seek?	This is frequently a neglected area; not mapping the concerned individuals or groups may provoke difficulties in this or later stages.
6	Opportunities	Internal opportunities. External opportunities (at the political level, in the context of public opinion, with partners).	
7	Risks	Internal risks External risks (at the political level, in the context of public opinion, with partners)	
8	Problem formulation	Summarize the key factors analyzed in a few sentences.	This formulation ("the problem") is fundamental and should be agreed upon by those involved in identifying subsequent solutions.

The fourth application involves a case of *decision-making in a technical domain.*[16]

For example, a private or public organization desiring to install a video surveillance system should respect the following six-step procedure:

1. Analysis of the risks.
2. Identification of needs.
3. Choice of technical devices.
4. Installation of the selected option.
5. Functioning of the installation (the organization will have to decide if it will first conduct a test, implement the whole system, and run in parallel the old and new systems).
6. Evaluation.

It is important to note that this six-step procedure should only be launched after an initial assessment of the problem aimed at ensuring the surveillance system works in the specific situation where it would be applied.

2.15 OTHER DECISION-MAKING MODELS

According to Dolan and Lamoureux[17] (1990), an individual may resort to three methods to make decisions:

- Intuition: Gladwell (2005) states that, "there can be as much value in the blink of an eye as in months of rational analysis." He argues that failing to trust intuition leads to errors. Intuition is for him a byproduct of our life experience, a sort of personal library in which our unconscious absorbs information and spontaneously regurgitates it when the need for it arises.[18] To support his thesis, Gladwell cites, among many others, the example of the Getty Kouros. This Greek statue at the Paul Getty Museum in California was proved by many experts to be legitimate but the head of the Acropolis Museum in Athens when he saw it said, "Anyone who has ever seen a sculpture coming out of the ground could tell that that thing has never been in the ground." Later it was proven that this expert's "blink of an eye moment" was correct. The statue was a fake.[19]
- Personal values.
- Judgment based on previous decisions in analogous situations.

---

16 Stéphane Leman-Langlois and Lucie Dupuis "Les technologies de protection des espaces" in M. Cusson, B. Dupont, F. Lemieux (eds), *Traité de sécurité intérieure*, Presses Polytechniques Universitaires Romandes, Lausanne, 2008, p. 438.

17 Dolan, S. and Lamoureux, G., *Initiation à la psychologie du travail*, Gaëtan Morin, Boucherville, 1990.

18 Gladwell, M., *Blink: The Power of Thinking without Thinking*, Little, Brown and Company, New-York, 2005, p. 74.

19 Ibid., 28.

One can identify another method:

- "Trial and error" which involves a strategy of trying to solve problems by proceeding with different forms of action until error is sufficiently reduced or eliminated.

2.16 SOME DECISION-MAKING TRAPS

One must be aware of the pitfalls and traps that can limit the rationality of the decision-making process – difficulties more apparent when a human element is involved than solely technical problems, where causes can often be signaled by the system itself. When it comes to "human" problems, symptoms can disguise causes (*e.g.* a decrease in the productivity of an individual could relate to family difficulties). Though decision-making is largely a rational process, it is a process bounded by different factors.

- *Physiological factors*: Physiological restraints consist of skills, habits, reflexes, the physical condition and psychological strength of a person. These elements together determine how individuals respond to certain situations.
- *Values*: Individual beliefs and values also influence the way people make decisions. One will not easily make a decision against his or her own values and beliefs.
- *Cultural factors*: These can have negative effects, for instance the mutual prejudices between uniformed and plain-clothes officers in a policing context can be an obstacle to sharing information.
- *Knowledge*: It is impossible to know everything and therefore one should realize that considering "the options" is not the same as considering *all* options; the amount of knowledge one can possess is limited.
- *Organizational factors*: Structures can hinder the flow of information or complicate interactions between actors.
- *Human factors*: Decision-makers are human beings and will never act as rationally as theory envisages. Individuals are not only subject to external pressures, but are also influenced by their personal interests, biases and prejudices. For example, political or other beliefs can prevail over professional skill when politicians recruit personal staff (indeed, the same can be true of directors, chiefs, etc). Attitudes towards ethnic minorities or religious believers can also be an important manifestation of prejudice; for example, France has tried to introduce professional resumés without names and photos to prevent candidates belonging to minority groups from being discriminated against on the basis of their name or skin color.
- *Cognitive bias*: Bias (resulting in errors of reasoning) has the potential to lead to catastrophe (see the Case Study below).
- *External factors*: Outside elements may influence the decision-maker in limiting his or her freedom and autonomy (e.g. changes in the political environment or crisis events).

Case Study: The crash of United Airlines Flight 173
While flying over Portland on 28th December 1978, the pilot of United Airlines Flight 173 heard puzzling sounds and noticed vibrations when the landing gear was operated. Believing the gear had not been deployed, the pilot began to circle over the region of Portland and initiated control procedures: passengers were prepared for an emergency landing. However, because all attention was focused on the landing gear procedures, nobody in the cockpit checked the fuel level. Ultimately, the plane crashed and 10 people were killed[20]; investigators at the scene of the wreckage found the landing gear had been deployed.

Table 2.37 regroups the main traps and pitfalls outlined into different stages of the decision-making process and suggest certain strategies for circumventing these obstacles.[21]

As mentioned previously, information plays a key role at every stage of the decision-making process. The phenomenon of "bounded awareness" must be taken into account here as it can represent an important obstacle to the efficient use of information. "Bounded awareness can be described as a situation when cognitive blinders prevent a person from seeing, seeking, using, or sharing highly relevant, easily accessible, and readily perceivable information during the decision-making process. Bounded awareness can occur at various points in the process. First, an executive may fail to see or seek out key information needed to make a sound decision. Second, they may fail to use the information that they do see because they are not aware of its relevance. Finally, executives may fail to share information with others, thereby bounding the organization's awareness."[22]

---

20 Christian Morel, *Les décisions absurdes*, Folios essais, Gallimard, Paris, 2002, chap. 1.

21 The article of John S. Hammond, Ralph L. Keeney and Howard Raiffa, "The Hidden Traps in Decision-Making", *Harvard Business Review*, September-October 1998, has provided elements for this section.

22 Max H. Bazerman and Dolly Chugh, "Decisions Without Blinders", *Harvard Business Review*, Special Issue January 2006.

Table 2.37 Decision making traps and pitfalls.

Traps and pitfalls	Reducing the risks of traps and pitfalls
Problem definition	
Potentially incomplete, imperfect, or even false information.	Cross-checking of information.
Being guided by first impressions.	Examining the problem from different perspectives; seeking other advice.
Accepting certain assumptions as true without further questioning (an example of a unchallenged strategic assumption was the firm belief that China would not cross the Yalu River in support of the North Korean government in 1950; the same type of assumption was made in 1940 by the French in disregarding the crossing of the Ardennes by German tanks).	Identifying and challenging these assumptions; if necessary adopting measures in order to improve procedures and controls.
Poor problem definition (erroneous or incomplete). A common error is to provide a problem statement that is plausible but incomplete in failing to account for all the consequences. For example, in the 1960s the American car industry defined the issue of road safety in terms of improving highways and better training drivers. The industry was taken aback when the government issued new regulations compelling technical improvements.[23]	Reformulate the problem from different perspectives, allowing sufficient time for this step in order to avoid premature decisions.
Settling on a scapegoat rather than the real causes of a problem.	Looking not only at individual actions but taking into account the broader problems such as the interactions of a group. The aim is to target root problems.
Inadequate communication.	To neglect to inform those who are concerned directly or indirectly with the problem or the decision.
Decision-making	
Analysis paralysis due to the accumulation of information for the sake of information.	Tailoring requests for information to the problem at hand; focusing only on the most critical information.
High degree of problem complexity.	Concentrating on three to four options only.
Lack of time.	Scheduling and careful time management.
Allowing personal bias or prejudices to influence actions.	Ensuring a participative decision-making process.

[23] Peter Drucker, "Une décision efficace" in: *La prise de décision*, Editions d'Organisation, Paris, 2002.

Table 2.37 Decision making traps and pitfalls (cont.).

Traps and pitfalls	Reducing the risks of traps and pitfalls
Decision-making	
Favoring arguments supporting one's own decision.	Ensuring a participative decision-making process; appointing a "devil's advocate."
Unconsciously or consciously favoring the status quo (including to prevent self-analysis).	Sticking to the decision-making process and listing other options to be examined.
Framing a new problem as the resurgence of an old one and resorting to known, but potentially inappropriate, solutions.	Allowing sufficient time for problem definition.
Placing too much confidence in forecasts.	Establishing ranges instead of clear cut figures; Developing different scenarios.
Group-think (after a certain time a group can develop a consensual approach and avoid disagreement).	Changing members of the group to keep its dynamic; using separate teams that evaluate two (or more) strongly-held, but competing, hypotheses.
Apply the same management style to each situation.	Adapting management techniques to the situation in question and to the maturity of collaborators; favoring a participative approach where possible.
Implementation	
Tasking the wrong individuals.	Matching the demands of the task to the profile of the collaborator.
Changing objectives during the implementation phase.	If the situation compels a change in objectives, considering the related consequences of that decision in order to maintain overall operational coherence.
Lack of control.	Developing feedback measures; managing by walking around[23] in order to keep in touch with the reality.
Assessment	
Unavailable – or inadequate – analysis of implementation results.	Developing debriefing procedures.
Hiding errors and/or justifying what has been done.	Seeking an assessment from an individual or group external to the project or operation.

24 MBWA (Managing by Walking Around) is an unstructured approach to hands-on, direct participation by the managers in the work-related affairs of their subordinates, in contrast to rigid and distant management.http://www.businessdictionary.com/definition/management-by-walking-around-MBWA.html (last access, October 26 2010).

2.17 TOOLS

Certain tools can be useful in the context of the decision-making process. Table 2.38 provides an overview of these instruments and methods, including in what context each can be employed during the different decision-making phases. Most of these tools have been developed, are used in a business management environment but they also provide useful instruments for police.

Table 2.38 Useful tools for the decision-making process.

Phases	Tools and methods
AWARENESS	Symptom identification (list of all visible manifestations, consequences and effects identified; one can begin to order symptoms into categories).
ANALYSIS Complexity. Identification of the causes of the problem. Statement of the problem.	Checking key assumptions. Collecting data on the possible background and effects of the problem. Interviewing individuals who can supply information on the subject. Brainstorming.* Scenarios and role plays* Analyzing data. Force-field analysis.* Ishikawa diagram (causes/effects).* SWOT.* PESTL.*
CONDITIONS Objectives. Constraints. Risks.	Brainstorming.
SOLUTION Possible solutions. Development of solutions.	Brainstorming. Force-field analysis.* Critiquing the decision: designation of a "devil's advocate"; use of separate teams to analyze the same options.
DECISION Evaluation of solutions Choice Decision test	Brainstorming. Elimination. Prioritization. Assessing in the context of objectives and constraints. Criteria matrix.
PREPARING FOR THE FUTURE	Scenario planning. "What if" analysis.
GENERIC METHOD	SARA.*

* Further description about these tools and methods is provided below.

2.17.1 Brainstorming

This method is very useful at different stages of the decision-taking process when employed in a structured way. Brainstorming allows one to identify problems, risks, consequences, possibilities and solutions in a timely manner (Table 2.39).

Table 2.39 How to structure a brainstorming session.

For maximum efficiency, a brainstorming session should involve some basic rules: • No more than eight to ten people should participate. • All participants should be treated equally. • No ideas should be considered naïve or stupid, nor be censored. • No ideas expressed should be subject to criticism, either verbally or through body language. • No questions should be asked *during* the session. • Participants should not come with pre-conceived ideas. • The number of ideas supplied is important.
Similarly, with regard to the structuring of the meeting: • The "rules of the game" should be set, explained and understood. • The question at hand should be clearly outlined. • Each participant should express one idea during his or her turn to speak. • Ideas should only be expressed orally, or in written form on a flip chart (one idea by post it). • Ideas should then be regrouped into clusters. • Associations of ideas should be encouraged. • Participants should ultimately select the ideas that deserve the most attention (tabulating votes).
Possible questions to be considered: • Who, What, When, Where, Why and How?

2.17.2 Cause and effect diagram (*Ishikawa* diagram)

An *Ishikawa* diagram (Figure 2.40) visually clusters diverse influences into related categories, helping to understand the relations between the different "families" of causes. A diagram of this type can also be used to structure discussion during a brainstorming session, or to later analyze the results of such activity.

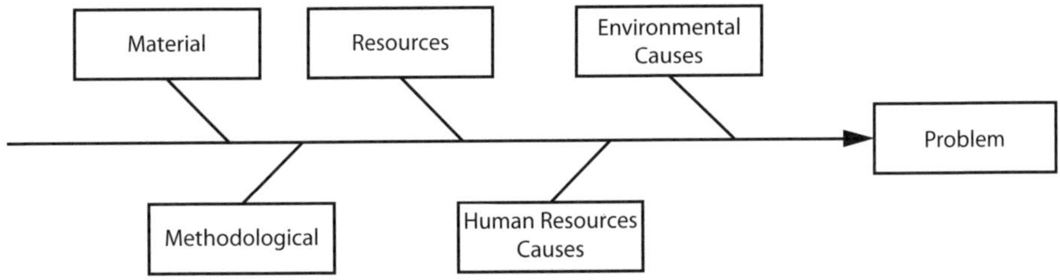

Figure 2.40 Ishikawa diagram.

2.17.3 Force-field analysis

In order to engage in this type of analysis, participants initially identify (list) the conflicting forces – positive or negative – influencing the outcome of a problem. In a second phase, these forces are ranked according to their influence. Finally, possibilities for capitalizing on the positive forces and reducing the negative ones are examined in order to arrive at a solution. Naturally, the brainstorming method can be used in this context.

2.17.4 The SARA methodology

SARA is frequently used in problem-oriented policing (POP), an approach focused on addressing the causes of problems of concern to the community and/or police. Effective measures are developed through cooperation between police and the community. SARA offers the methodological framework for structuring such cooperation: the four steps involved (Scanning, Analysis, Response and Assessment) are in fact derivative of the basic rational model, although at the same time tailored to the specific problem at hand.

- *Scanning* aims at identifying the consequences of recurrent problems and selecting those requiring closer examination.
- *Analysis* deals with defining the problem (identification of relevant data is a key element).
- *Response* refers to the listing of options, a choice between them, and the articulation of an action plan.
- *Assessment* involves the reviewing of the process, data and results.

The Royal Canadian Mounted Police (RCMP) has adapted and refined the method, employing a variation labeled “CAPRA,” where more emphasis is placed on Clients and Partnerships.

2.17.5 SWOT

A SWOT analysis is a strategic planning tool used to identify the strengths, weaknesses, opportunities and threats *for assessing an organization, a project, an operation or a business venture* (Figure 2.41).

- *Strengths* are attributes of an organization that contribute to achieving objectives and/or exploiting new opportunities.
- *Weaknesses* are attributes of an organization that are harmful to achieving objectives or completing projects.
- *Opportunities* arise as a result of shifts in an organization’s external environment, which open new possibilities for development or present favorable conditions for activities.
- *Threats* are consequences of shifts in an organization’s external environment, which can have an adverse influence on projects or activities.

	Facilitates achievement	Obstacles to achieving objectives
Internal Factors	**Strengths**	**Weaknesses**
External Factors	**Opportunities**	**Threats**

Figure 2.41 SWOT grid.

Opportunities and threats can be assessed through brainstorming sessions, or by using the PESTL method described below.

2.17.6 Scenarios

Scenario planning is a method for preparing for future activities by understanding the nature and impact of the most uncertain and important driving forces affecting an organization. It is a group process centered around the goal of crafting a number of diverging stories through an extrapolation from the existing state of knowledge.

The following steps can be useful for the development of scenarios:

- Define the issue to be addressed.
- Identify major stakeholders.
- Map basic trends and driving forces (see the use of the PESTL method on the next page).
- Identify key uncertainties, events that can/cannot occur (only those that are relevant, significant and unpredictable should be taken into account, and linkages between driving forces should be identified).
- Identify the possible outcomes of the driving forces (likely, least likely, unlikely).
- Construct three scenarios (worst, best, likely).
- Identify the issues arising and examine where measures should be taken/assess where it is possible to take measures.
- Assess the scenarios (are they relevant, internally consistent?).
- Identify research needs and assess where further information is needed.

Scenario planning can be used in a number of ways – such as for environmental analysis, for defining a strategy, or for establishing the framework for a simulation exercise.

In an operational context, consistent analysis of a situation in order to determine possible developments (what an adversary could do or how an environment could evolve) is imperative. Such hypotheses form the basis for the establishment of contingency plans or of counter-measures.

Case Study: Using scenario planning

- A consultant was involved in a security assessment of a large company that received, warehoused and shipped valuables. In order to identify risks, vulnerabilities and develop solutions, he used the following procedure:
- A brainstorming session was conducted to list and rank the security risks faced by the company. Participants (managers of different levels) were asked to put themselves in the shoes of offenders planning to steal valuables (including paintings, jewelry and precious statuettes) and to identify in broader terms the general risks.
- On that basis, the four most dangerous and likely actions were selected.
- Four teams were formed; each was required to develop a scenario for realizing one of the four threats.
- Finally, war games were organized in which one team put into action its scenario and the others played the role of defenders. The exercise facilitated the assessment of security measures, identification of vulnerabilities, and subsequent discussion of potential risk management measures.

2.17.7 PESTL

The PESTL method can be employed to generate a useful analysis grid focused on a particular environment, while also contributing to the preparation of scenarios. PESTL reflects a perspective oriented around the different elements of a system: political, economic, social, technological and legal. Data must be collected and analyzed in each of these categories to identify changes and their consequences for the organization. This will enable one to define, in comparison with the state of the organization (through the SWOT method described above) what threats and opportunities such developments may encompass.

Case Study: Using PESTL

A police commander employed the grid below in annual brainstorming sessions to identify and assess changes in the environment that could *influence* the police force. For example, new elections could *affect* relations with relevant political authorities. The PESTL method was used to determine what consequences could arise and what could be done to minimize any negative effects or capitalize on any new opportunities (*e.g.* establishing contacts with the potential new political leader, preparing an information folder for facilitating the entry into his or her new position, etc).The same approach can applied to assess legislative changes.

One important element of success is to define as precisely as possible a list of representative indicators to follow (Table 2.42). PESTL is very useful in fostering a participative style of management; brainstorming proved an efficient method for generating material for the different categories (though supplemented by other approaches such as desk researches, interviews, etc.).

Table 2.42 A PESTIL grid.

Sectors	Changes	Consequences For the organization	Threats	Opportunities	Autonomy of action	Possible Measures
Political						
Economic						
Social						
Technology						
Legal						

2.18 INTEGRATION OF INFORMATION IN THE DECISION-MAKING PROCESS

As already developed in Chapters 1 and 2, information spans all phases of the decision-making process – from helping decision-makers to become aware of problems, to supporting preventive measures when necessary, decision-making, monitoring and preparations for the future. Figure 2.43 illustrates the different decision-making levels and relevant time frame.

There are different levels at which information is collected and analyzed. Three layers are generally considered either on a conceptual basis: strategic, operational, and tactical or on a geographic basis: national, regional, and local. Naturally, these two dimensions can overlap, as the strategic analysis will be done at the national level, the operational analysis at the regional and the tactical analysis at the local. The development of the transnational cooperation today adds a fourth level, the international one.

Table 2.44 summarizes the links between decision-making and information at different levels.

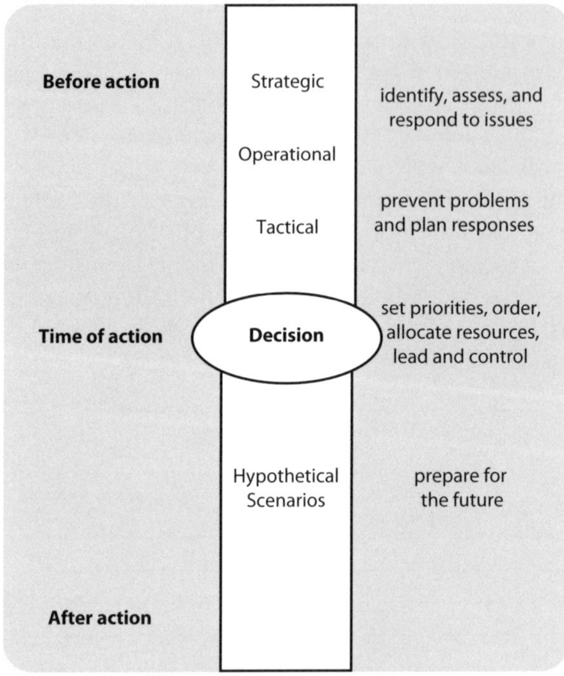

Figure 2.43 Levels and time frame.

Table 2.44 Decision making and information levels.

Level	Action	Data	Method	Output
Strategic (national)	Scanning of the environment.	List of indicators in each category of PESTL.	PESTL, list of indicators.	Opportunities and threats.
	Strengths and weaknesses of the organization.	Necessary for assessments.	SWOT	Capabilities of the organization to exploit the opportunities and withstand the threats.
	Objectives and measures.			Outline of a strategy and plan of action.
Operational (regional)	Necessary operations to achieve strategic aims.	To assess, monitor, evaluate.	Project coordination/operational management according to decision-making methods.	Operational orders.
Tactical (local)	Necessary operations to achieve operational aims.	To control and monitor the actions taken.		

At a strategic level, it is important to scan the environment in order to understand any consequential shifts; this involves identifying new threats and opportunities and evaluating if, and how, they could affect the organization in question.

Case Study: Canada's strategic warning and intelligence project[25]

Warning is described as "the product of an intelligence judgment on the level of threat and risk posed by a particular enemy or scenario. Most important however, warning must be communicated; an analytical judgment only becomes a warning when it is communicated and understood as such by the receiver." Warning begins with the presumption that a potential threat exists in order to develop an indicator list, which will be used to follow the possible development of the identified threat. This military system has been adapted for fighting organized crime. The process for the development of warning intelligence mirrors the traditional intelligence cycle with one key exception: its inception. The strategic warning is a case of bottom-up intelligence – the process is initiated by the analyst, and the finished intelligence is moved upwards to senior decision-makers.

2.18.1 Direction

Data collection must help leaders assess a situation and make sound decisions. As time and resources are generally scarce, the gathering of information must focus on what is important in relation to problems faced. For this reason, the direction provided by leaders for the search for information is critical. The formula 5W + H can be used to define the data to be collected and analyzed (Table 2.45).

Table 2.45 Use of 5W + H to define data collection and analysis.

WHY	Purposes.
WHAT	Kind of information to be collected/priorities.
WHO BY WHOM	Sources; Resources.
WHERE	Area(s) of collection.
WHEN	Period of collection/rules of transmission.
HOW	Methods, resources.

2.18.2 Order for information and collection plan

The Chief of Operation (or the intelligence officer if the structures established are important) must state clearly his or her request for data in an order for information (see also Sect. 1.4.3):

[25] "The Strategic Warning for Criminal Intelligence," Daniel R. Morris and Gregory Baudin-O'Hayon, *International Handbook on Risk Analysis and Management*, Center For Security Studies, ETH, Zurich, 2008.

- *Routine* information to be collected is outlined in a document that remains valid irrespective of modification or cancellation. The document lists information of a generic nature, that is, not directly related to the nature of the problem at hand.
- *Specific* information is requested as part of a separate order linked to a particular case or situation.
- *Key* information[26] is indicated.

2.18.3 General model

The general model represents an example of information collected according to the 5W + H formula to support the planning of an event; the model also shows how this information is subsequently used to underpin specific action (Table 2.46).

Table 2.46 Use of 5W+H formula applied to the general model.

Areas	Information	Measures
WHAT	Types of threats.	Prioritization and deployment of resources.
HOW	Possible actions by the offender possible evolution of the situation, and modus operandi.	Contingency planning, countermeasures.
WHO	Possible contributors. Possible targets.	Preventive measures. Levels of protection.
WHEN	Timing of possible actions.	Timely deployment of resources.
WHERE	At which places.	Geographical deployment of resources.
WHY	Purposes.	Planning.

The template of the *Order for the Search for Information* featured in Figure 2.47 is closely linked to the use of the Tables presented above.

ORDER FOR THE SEARCH FOR INFORMATION CLASSIFICATION

CHIEF OF OPERATION or INTELLIGENCE OFFICER
CODE OF OPERATION

ORDER FOR THE SEARCH FOR INFORMATION

1. PRIORITIES FOR THE SEARCH FOR INFORMATION

First: For instance: number of participants, offensive equipment, intentions; Practicability of the roads leading to and in the area of the demonstration.
Second: For instance: incidents in surrounding areas.
....

Figure 2.47 The template of the *Order for the Search for Information* (cont. on p. 80).

[26] Key information is information that must be forwarded immediately, without following the whole chain of command, to the chief of operation as it can have a decisive influence on his decision (see 1.4.3).

2. TYPES OF INFORMATION

2.1 Key information

CODE	KEY INFORMATION	SOURCES	TRANSMISSION
If necessary a code name is given for key information.	Information thought to be key to the decision-making of the Chief of Operation. For instance: • Departure of the demonstrators from their gathering place; • Arrival of the demonstrators in the sector X; • Assaults on security personnel.	Who is the provider of this information?	Chain of transmission defined. Key information must generally be transmitted directly to the Chief of Operation.

2.2 Specific information

CODE	INFORMATION	RESEARCH AREA	SOURCES	TRANSMISSION
If necessary.	Information to be specifically collected for a particular operation. For instance: • Situation at areas X, Y, Z; • Number of participants.	Generally this information is related to specific areas that must be indicated.	Who are the providers?	Rules of transmission, for example each hour to the intelligence officer.

2.3 Routine information

General information, not specifically related to an area or even to the operation that can be useful for an assessment of the situation. This information must be specified and provided by all sources in all areas.

For instance:

- Placement of blockades;
- Violent incidents;
- Attitude of the population;
- Practicability of roads.

3 COLLABORATION

3.1 Reinforcements

Indication on supplementary resources, if any, which are attributed.

3.2 Exchange of information

Partners must be indicated and, if necessary, restrictions in the exchange of information should be made clear.

4 SPECIFIC INSTRUCTIONS

- Special equipment;
- Communications-radio-telephone, prearranged, signals;
- Others.

5 LOCATIONS

Figure 2.47 The template of the *Order for the Search for Information* (cont.).

2.19 SUMMARY AND KEY POINTS

- Different decision-making methods exist. The rational model offers many advantages – in particular, the model deals systematically with the four steps of the decision-making process and hence serves as a broad framework applicable in different situations.
- The rational model can be applied in different situations even if certain specificities will depend on the context of its application (illustrated by the examples presented).
- The decision-making method described cannot only assist in problem solving, but also in organizing work, presenting options and writing reports.
- One must be aware of the potential traps and pitfalls in the decision-making process, with several possibilities for reducing proposed risks.
- At each stage of the decision-making process, specific methods may be applied. Though the intelligence cycle is used mainly in the first phase, information is integrated throughout and adapted to the particular requirements of each step.
- The efficiency of the process rests in large part on the management of information, which should be guided by the following features:
  - Clearly stated requests for information.
  - Distinct prioritization in order to avoid too great a dispersion of effort and resources.
  - A focus on the timely transmission, rather than interpretation, of collected information.
  - The identification and use of all sources able to supply required information.
- The following information-related risks must be reduced:
  - Requests for information that lack clarity.
  - Too narrow a use of sources.
  - No clear identification about the recipient of the information.
  - Confusion between routine and specific information.

THIRD PART OF THE DECISION-MAKING OPERATIONAL CHAIN: *TACTICS AND DEPLOYMENT OF RESOURCES*

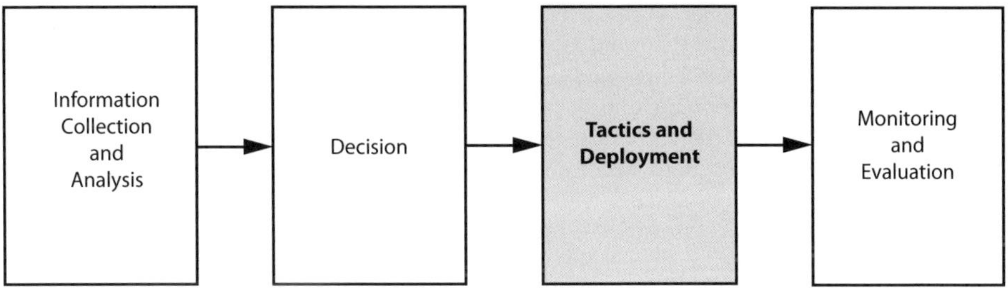

Figure 2.48 Tactics and deployment is the third stage of the decision chain.

Objectives

This Section presents an overview of the principal police tactics, examines some tactical considerations, discusses how to increase the efficiency of resources and reviews some of the issues regarding the choice of a specific tactic.

2.20 WHAT DOES TACTIC MEAN?

While the term "tactic" has many meanings, for the purposes of this book it shall be used in connection with the tripartite ends, ways and means framework described in Chapter 1. Tactics thus represent the ways to achieve a goal defined by a strategy. A strategy can be understood broadly as the general direction taken to meet overall objectives; tactics belong to the execution phase where a strategy is implemented by short-term decisions linked to specific actions. A strategy is largely determined by the available means, while tactics involve using such means in the most efficient manner. Tactics and means are therefore strongly interdependent.

2.21 A FRAMEWORK FOR POLICE TACTICS

The following framework outlines the tactics that can be used to achieve goals at a strategic, operational, tactical or individual level (Table 2.49). The tactics may serve preventive as well as repressive purposes.

Table 2.49 A framework for police tactics.

Level	Definition	Types of tactics[27]
Policies and strategies.	Long term strategy (at the police level) to fight crime, focused around a new paradigm of policing that implies changes in methods, organization, training, and interaction with other actors in the security sector. General objectives: to better prevent or fight crime in partnership with other security system stakeholders and/or society. These strategies will use different tactics.	One should speak here of policing models (broad concepts) that use a mix of tactics: • Problem oriented policing; • Community policing; • Intelligence-led policing; • Zero-tolerance policing;
Operational.	Tactics employed at the police level, or more generally within a specific region, to support planned operations addressing serious, permanent and clearly defined problems in space and time.[28] General objectives: to suppress or diminish crime phenomena in specific areas and avoid troubles during large events.	• Crackdowns; • Crowd control; • Large event security; • Risk-based policing; • Campaigns against impaired driving; • Targeting of prolific and serious offenders.
Tactical.[29]	Tactics used in the deployment of patrols or of small units to meet day-to-day operational requirements as well as specific incidents that may occur. General objectives: to deter and prevent crime.	• Random patrol; • High visibility patrol; • Low visibility patrol; • Directed deterrent patrol; • Target-oriented patrol; • Split-force patrol; • SWAT teams;
	Tactics used in emergency cases to pacify dangerous individuals.	• Emergency teams, negotiation teams; • Bomb squads; • Defensive and offensive driving;
	Police vehicle tactics.	• Pursuit driving; • Barricades and roadblocks; • Vehicle stops, routine and high risk; • Scene approach and security;
	Crime scenes.	• Scene management; • Identification: witness interviews, forensic analysis;
	Investigation tactics.	• Location: surveillance; • Structuring evidence.
Individual.	Defensive and offensive tactics used by individual police officers to accomplish their missions and ensure their safety.	Hand to hand combat. Shooting, baton, pepper spray and handcuffing.

27 These tactics are detailed in the Annex of this Chapter.

28 Maurice Cusson, Benoit Dupont and Frédéric Lemieux (eds), *Traité de sécurité intérieure*, *op. cit.*, p. 131.

29 Charles D. Hale Police Patrol, Operations and Management, Prentice Hall, Upper Saddle River, Pearson, 2004, pp. 360ff.

2.22 TACTICAL CONSIDERATIONS

Four principles must be taken into account when the choice of a tactic is discussed for deploying resources (Table 2.50). Acknowledging these principles will assist the operational chief during the decision-making process.

Table 2.50 Principles behind the selection of a tactic.

Principles	Developments
Freedom of action (flexibility).	The operational chief must retain the capacity to influence an ongoing operation; this depends especially on the establishment of forward planning and the availability of reserves to react to new developments.
Concentration of resources on one point.	Military history teaches that a stronger enemy can be defeated by concentrating forces in key sectors. The concentration of police forces in hot spots identified by intelligence (targeting) can produce excellent results.
Adequate allocation of resources.	An adequate allocation of resources enhances the capacity to use them efficiently. In that context, a sound balance between specialists and generalists is important.
Indirect approach or diversity of action.	The strengthening of cooperation and synergies with partners within the security system and civil society helps to boost the effectiveness of operations and can contribute to a reduction in problems.

Case study: British principles of war

- Selection and maintenance of the aim (the commander should only have one aim at any time, and it should determine all actions).
- Maintenance of morale (partly sustained through a common sense of purpose).
- Offensive action (a state of mind conferring the initiative on the attacker).
- The element of surprise.
- Security (reduces vulnerability to enemy attack, requires only calculated risks).
- Concentration of forces (having more than the adversary at the decisive time and place).
- Economy of effort (judicious and cost-effective expenditure on resources).
- Flexibility (ability to adapt to unpredictability of military operations).
- Cooperation (team spirit of all units involved).
- Administration (exemplary staff work, efficient logistics).

2.23 COHERENCE

The coherence triangle featured in Chapter 1 showed how important it is to establish coherence between ends, ways and means (Figure 2.51). At the same time, it was noted that there is not a right and wrong approach but that the process is iterative. A tactic must be chosen, *inter alia*, on the basis of available means, not only in a quantitative, but also qualitative sense. Thus, resources represent a constraint.

Consequently, flexible resources enable a decision-maker to attain wider operational options. Hence, a Chief of Police must aim at increasing his or her freedom of action by creating the conditions for acquiring this flexibility. How can such a result be achieved?

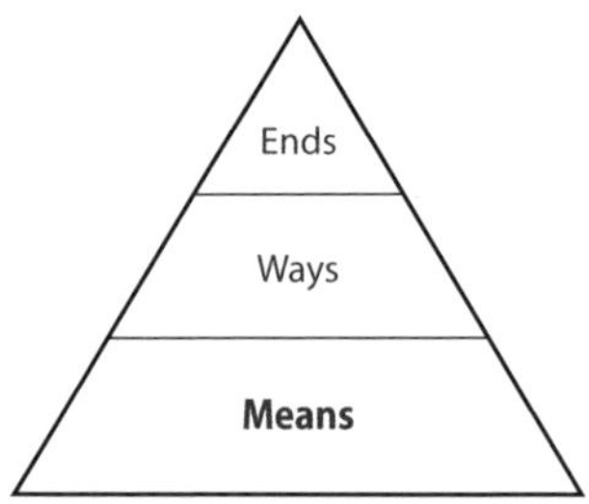

Figure 2.51 Triangle diagram.

2.24 HOW TO INCREASE FLEXIBILITY IN THE USE OF RESOURCES?

One must first understand what can limit flexibility in the use of resources. The main hindering factors related to this question are summarized in Table 2.52. Just after the Table, we will consider how these factors can be overcome.

Table 2.52 Limitations of flexibility of resources.

Availability of skills.	The necessary skills are not present; in most police forces, for example not all police officers are trained for crowd control interventions.
Availability of authority.	The necessary authority (*e.g.* legal power) is not available; there may be personnel within a police force that do not have the necessary powers to execute certain tasks.
Availability of personnel.	The number of personnel available is not sufficient in relation to the tasks.
Availability of time.	Resources are not, or only partly, available for the duration of an operation; this is the case when reinforcements are only available for a limited period, inadequate for the duration of an event. This may also be the case during the night or on weekends when the number of active police officers is reduced.
Geographic availability.	The necessary resources are not, or only partly, available in the area of operation.
Shortage or inadequacy of equipment.	Such factors can reduce the number of personnel able to accomplish certain types of missions.

2.24.1 Availability of skills

One logical approach to a dearth of necessary skills is to enhance the flexibility of police personnel by reducing specialization and increasing the number of generalists. If specialization has many advantages centered on better efficiency and effectiveness in large departments, one must also be aware of the key disadvantage: introducing greater organizational complexity and creating difficulties in communication and coordination. Specialization also increases competition and envy between different branches of a force, for instance between patrol officers and detectives. The relations between field officers[30] and non-field staff may also grow tense as the number of staff expands to meet the needs of coordination arising from increased specialization. Another unfortunate consequence of specialization in a police force is to greatly reduce the number of officers available for general duties. In all large police departments, there is a clear distinction between uniformed police officers and plain-clothes detectives. The latter are not only equipped differently, they may not have received the same training. Consequently, they cannot be integrated into riot-control units. In other forces, the lack of personnel for certain operations may be less acute, at least in theory, if all officers have gone through the same basic training and thus can be deployed in uniform if necessary.

Specialization may also be seen in geographical terms: Cusson[31] differentiates between the general surveillance exercised by police in a territory and planned operations where police address a specific problem. For instance, crackdown operations that concentrate forces during a certain period in a specific area in order to repress drug trafficking. This type of operation produces better results than general surveillance, but is frequently hampered by the lack of available officers. Hence, one is led to the question of the best balance between police officers tasked for general surveillance versus police officers devoted to planned operations. The research undertaken by Greenwood, Chaiken and Petersilia in 1977[32] provoked lively discussions within police circles in showing that at least half of the tasks accomplished by investigators could be dispensed with without consequence, as most cases are solved through information provided by witnesses. Patrol officers generally arrive first at a scene, their role in the identification of perpetrators is crucial as they conduct the preliminary interviews of individuals able to provide information. Thus, the question is whether it would not be more efficient to increase patrol officers and reduce the number of investigators. The same dilemma may be found in the balance between community police officers and other members of the force.

---

30 Field officers refer to those officers who are in direct contact in a case or with a problem.

31 Maurice Cusson "Questions de stratégie pour la police", in: *Traité de sécurité intérieure, op. cit.* p. 130.

32 Greenwood Peter W, Jan Chaiken, John Petersilia, Linda Prusoff, *The Criminal Investigation Process*, Heath and Company, Lexington, 1977.

Case Study: Community police officers and intervention patrols

The experience in the Canton de Vaud (Switzerland) illustrates the importance of the question of the best ratio between different divisions of a police force. In the context of a merger project between the Canton's police and municipal forces – a project that generated strong opposition amongst the municipalities and local police forces – three test areas were chosen where all intervention patrols were to be managed by the Canton and the community police officers by the municipalities. In order to achieve the intervention standards established (four minutes within the localities and twenty minutes within the state), the number of patrol cars had to be increased and several municipal police officers integrated into the Canton's mobile force. Local political authorities soon complained there were fewer police officers on the beat and the population felt insecure. Moreover, the authorities claimed the security level had decreased – though no evidence demonstrated this – and the merger project was abandoned, meaning the maintenance of several police forces continued to hamper general efficiency.

The choice between generalists and specialists is a difficult one and there is no right or simple answer. If generalists are more apt to be engaged in diverse operations, their level of knowledge can be too low for certain tasks. Training cannot make up for all deficiencies; the investment required to bring the whole force to the same level across all types of activities would be very unrealistic. Many tasks can only be mastered by specialists. The problem is the number of such tasks tends to be multiplied, as police work grows increasingly complex. This trend has adverse consequences for the organization and management of a force. Each new criminal development leads to tailor-made responses and frequently to the creation of task forces or specialized units. As one cannot address continuous phenomena with temporary structures, these task forces and units are inevitably made permanent, with perverse organizational, managerial and financial effects. A permanent structure will involve a permanent manager with an assistant, and possibly a secretary, office premises and new equipment. Hence, the effect on the overall manpower of the force will soon be felt in other areas of the organization if no new staff is recruited. The chief will have difficulties in controlling such an evolution, as officers will strongly support the new possibilities opened for promotion. In organizational terms, beyond the loss of flexibility induced by the creation of a new specialized unit, the structural fragmentation will have consequences for the flow of information and the relations between different areas of the force: the envy between units may increase and local fiefdoms could appear at the expense of organizational unity.

Nonetheless, one should not conclude that specialization is inherently bad; indeed, specialization may be a necessity. An organization made up of generalists would soon reveal its weaknesses. In a complex environment such as that faced by police forces, such a system would not be efficient. The solution must be pragmatic; one must analyze where specialization is necessary and limit it to certain areas. The danger is in creating specialized positions when and where they are not necessary. For example, consider the following two case studies.

Case Study: Road traffic police

In Switzerland, the most significant difficulty for police in the 1960s was related to vehicular traffic. Increased traffic and the construction of highways led to the development of a new branch, the traffic police. To mark their differences, the traffic police officers, who were driving powerful cars, adopted a different uniform (big black boots were one of its characteristics). Within the police force a different culture appeared. Traffic police officers considered themselves as the aristocrats of the force and regarded with contempt the other uniformed police officers. In retaliation, the latter did not spare their criticisms of the former. Tensions were sometimes exacerbated to the point that traffic police officers were sometimes forbidden to enter police stations. The sharp increase in crime during the mid-1970s led to a reappraisal of the situation. Chiefs of police realized the separation of the uniformed police into two discrete branches was too expensive in financial terms, and in terms of availability of resources. Consequently, the traffic police units either merged with other units or expanded their areas of responsibilities notably in the field of surveillance. Specialized personnel were kept for very specific tasks such as radar, and truck control. The result was largely successful, as tensions were reduced within the force and the number of personnel available for crime fighting increased.

Case Study: Stolen vehicles squad

In one Swiss police force in the 1990s, analysis of relevant data pointed to the economic impact of vehicle theft. The analysis showed the "income" produced by this type of crime was larger than the proceeds resulting from burglaries. A task force was set up to address this problem and produced very quick results. Several gangs were dismantled and the situation returned to a normal level. Consequently, the *ad hoc* task force was disbanded and its members returned to their former units. These officers were not happy with the decision, as they had hoped to remain members of a new unit that would have offered new promotion opportunities.

What measures could potentially limit the disadvantages outlined? Based on the examples given in the two case studies above, one might take one or more of the following actions.

- Maintain a sound balance between specialists and generalists – specialized positions should only be created when a specific need for expertise cannot otherwise be met.
- Assess the utility of an activity and, if appropriate, consider termination or outsourcing.
- Simplify organizational and hierarchical structures wherever possible, in order to reduce the organizational fragmentation.

- Ensure a system of rotation between units in order to avoid the development of organizational subcultures.
- Refrain from transforming task forces into permanent units.
- Conduct regular organizational assessments to ensure that necessary changes are completed.

2.24.2 Availability of authority

Problems can arise when a police officer who intervenes in an incident is not sufficiently empowered to take necessary action (*e.g.* does not have the legal authority to take certain actions). This is especially the case when different police forces operate on the same territory, with different legal powers. Investigators may also enjoy particular powers other police officers do not. For example, in France there are three main police forces: the National Police, the Gendarmerie and the Municipal Police. Only personnel serving with the first two forces have full police law-enforcement powers. Members of the municipal police forces do not have authority to conduct most law-enforcement measures. In Switzerland – a federal state – different police forces exist in several cantons (states). With few exceptions, only members of a Canton's police force have the authority to proceed to criminal operations. Situations where some police officers enjoy full police powers, and where others do not, creates problems. The most common problem relates to victims' frustrations when faced with officers who lack the necessary authority to act. The following example illustrates the difficulties caused by such situations.

Case Study: Where should a complaint be filed?

One evening, around 8:00 PM, two young girls in a small town near Lausanne (Switzerland) were robbed. They went to the city police station where they were the receptionist explained to them that he was unable to register their complaint as the city police were not authorized to do so. He directed the two girls to the canton precinct located in the same town. The precinct was closed as the regional patrol center assumed responsibility after 6:00 PM. The girls decided to go to the nearby city police of Lausanne. Here, too, the police could not process their complaint as the robbery occurred outside the force's jurisdiction. The girls were sent to the canton police station in the town, which was also closed. They noticed, however, that a light was on in one window of the building. Climbing the staircase, they knocked at the door. A man opened the door, listened and sympathized with their situation, but expressed regret that he was also not able to do anything. This man was the minister in charge of the security of the canton but was not authorized to handle their complaint. Ultimately, the minister arranged for the girls to be driven to the canton police headquarters where at last their complaint was received. This anecdote is true and illustrates the problems created when different police forces, with different powers, coexist on the same territory. Any capacity to flexibly deploy resources is inevitably diminished in such circumstances.

Different solutions have been proposed – or even implemented – in order to remedy this counteractive situation. Some municipal authorities presented a request to the Canton in order to achieve greater powers for their local forces to deal with issues of this nature… However, such solutions would only increase confusion across the police community and enhance competition between forces. Unconstructive competition of this kind can lead to hoarding of information, complex coordination measures, or clashes of egos. All such factors hinder police efficiency.

Case Study: Olympic games, Atlanta 1996

Security headquarters were established in a large Atlanta hotel the week before the opening of the 1996 summer games. A modern functional operational room had been set up and each agency had a well-equipped work area. Many agencies were represented: from hotel security, to town police, the sheriff department, state police, federal agencies, the FBI, marshals and the secret service, as well as the National Guard and the regular army. The question "who is in charge?" was answered by "it depends on the case." For instance, a theft in the hotel would be handled by the hotel security, while a murder would be investigated by the city police. However, if an out-of-state perpetrator committed the murder, the FBI would be in charge. In such a situation, inevitable problems will arise over identifying a lead agency when faced with a complex case, with valuable time potentially lost in the process. This was the case when a bomb exploded one night in Centennial Park a few days after the games had begun. According to well-informed sources, the question of who was in charge of informing the media seemed to have been a major concern and took at least two hours of lively discussion to solve.

Case Study: Parking a police car in an American city

On a more humorous note, while at the FBI National Academy at Quantico, a Swiss officer went with some American friends to a nearby city where it was difficult to find a parking space. Seeing a place reserved for police, the officer suggested the group should take it. The driver responded, "Impossible, if they know we come from the FBI, they will double the fine."

Which measures could reduce such inconvenience?

- The extreme solution, although difficult to achieve politically, would be to reduce the number of police forces operating in a given territory. Ideally, there should be only one, whose members would all enjoy the same authority. This is achievable in merging differences forces. Such a move has been realized in certain Swiss cantons but in others, it has failed; in Canada, much progress has been made through police regionalization efforts.

- If such a development is impossible, the drawbacks of an atomized organization can be reduced if legally, it is clearly established that the main police force (reporting to the highest political level in the territory) is entitled to direct and coordinate the other forces whenever necessary. This is particularly pertinent when large events necessitate the deployment of a large contingent of police officers drawn from different forces or when cases – e.g. a series of burglaries – concern different political territories and must be centrally managed. Unless clear rules are established, municipal authorities may be reluctant to place a portion of their force under the command of the canton, state, or federal police, therefore hampering operational efficiency. The tendency to set up operational or liaison committees in order to facilitate cooperation or reduce interpersonal rivalry (such as clash of egos or mutual prejudices belonging to different entities) does not foster efficiency, as such groups are in general not suited to rapid and clear decision-making.

- One possible system for clarifying relations and interactions between different independent police forces is connected to the introduction of "Operational Management by Objectives" (OMBO). Such a system can naturally only be set up when all the concerned parties agree to cooperate. Objectives are fixed according to information collected and analyzed and should establish the basis for deploying resources. In the context of autonomous police forces that have agreed to institute better coordination mechanisms, the objectives should contribute to tasking and the rules of collaboration. A possible system might be organized according to Table 2.53.

- Increased cooperation between different Forces can also be a solution to resources, expertise and cost issues, while at the same time increasing efficiency and effectiveness across jurisdiction areas, as underscored by the case study on the Integrated Homicide Investigation Team (IHIT) presented below.

Table 2.53 A system for OMBO.

Type of objective	Definition	Who shapes the objective?	Operational-chief	Priorities
Strategic/ National	Concerning operations falling within a general strategy, or covering an entire territory (or many of its jurisdictions).	Political authorities, in consultation with the Chief of the State Police, using a variety of tools such as community consultation, environmental scanning and the analysis of criminal intelligence.	Chief of Police.	These objectives are overarching and involve mandatory implementation and reporting.
Operational/ Regional	Covering several jurisdictions in a clearly identified region; regional resources are sufficient and if necessary reinforcement can be authorized by senior managers.	Regional chief, if necessary in collaboration with regional authorities using the same tools as above.	Regional Chief.	These objectives are aligned with strategic objectives whenever possible and have precedence over local objectives; progress reporting is provided to all.
Tactical/ Local	Concerning operations conducted on the territory of a single jurisdiction.	Police and local authorities using the same tools as above.	Local Chief.	Reinforcement may be requested from the regional level, while the upper level is kept informed.

Case study: Integrated Homicide Investigation Team (IHIT)

The Integrated Homicide Investigation Team (IHIT) is responsible for investigating homicides, police-involved shootings and in-custody deaths occurring within the British Columbia Lower Mainland areas policed by the RCMP, Abbotsford, New Westminster and Port Moody police departments. The IHIT is a truly integrated unit in that four police agencies work together to conduct investigations. One of the advantages of integration is that experts in certain fields of investigation can be tasked to deal with specific problems. Integration also allows partner agencies to quickly deploy a full team of specialized investigators to any homicide. The IHIT currently has a total of 109 employees, of whom 84 are investigators or detectives.

Case Study: Canada's use of the Risk Management Process to assist in identifying strategic priorities

The Royal Canadian Mounted Police (RCMP) is utilizing the Risk Management Process to assist in identifying strategic priorities. Priorities are shaped by many factors and actors, such as government or unforeseen circumstances. However, for the most part, the process of priority setting is based on risk analysis.

While the concept is the same, the approach differs slightly across the various RCMP jurisdictions. For example, in the Province of Saskatchewan, officers collect and analyze data from many sources, such as the police detachment or station plans and unit quality assurances, a formal structure of self-audits, local and regional threat assessments, environmental scans, and audit and review reports. Risk registers are constructed in accordance with law enforcement and management activities at the local and regional level, identifying risks faced. Where a certain level cannot manage a risk, the issue is elevated to the next level until it reaches the top of the organization within the Province. At this stage, the risk register of the Saskatchewan Commanding Officer is then used as part of the strategic planning and priority setting for a balance score card (the performance management tool used); the risk register is also used to develop performance agreements with managers, as well as identify audit and review processes. Each involves a separate reporting procedure and timeline, which together feed back into the next cycle of risk assessment.

In order to manage, monitor, report on and lead this process, a dedicated operations strategy branch is essential. The branch should be structured to include communications, strategic support, business support and operational support elements. Communications personnel ensure all aspects of internal and external communication activities are addressed. Strategic support comprised of research analysts and planners assists with the risk assessment process and, once mitigation strategies are identified, integrates them where appropriate into planning and reporting tools. Strategic support personnel also assist in developing strategies, subsequent evaluation and interpretation of results and look for best practices and empirical data to ensure strategies are evidence based. Business support personnel provide support to all with respect to financial management, including direct assistance to front line police managers. Operational support personnel provide direct support to detachment and stations and regional commanders in fulfilling their policing obligations to local authorities, human resourcing and equipment requirements.

2.24.3 Availability of personnel

Adequate resources may not be marshaled to face a complex situation, or, resources could only be available for a limited timeframe. Complications may also arise from the fact that forces are dispersed between different jurisdictions or political territories. Examples of such situations may be found when a large event takes place in a country; for instance, the Olympic Games or at a meeting of the G20. On a smaller scale, many Friday and Saturday nights see increased belligerent activities, while football matches are subject to hooliganism. In many cases, local forces cannot independently address such problems and must receive reinforcement from other forces. This reinforcement cannot always be provided, or at least not for the duration required.

We now consider measures that are capable of reducing such inconveniences. Four types of measures – targeting, use of reserves, gradual reinforcement and partnerships – offer the best means of addressing such issues. These four measures can often be combined.

Targeting

Targeting in this context involves identifying levels of risk and deploying forces accordingly.

Tables 2.54 and 2.55 are a simplified example of a system that was used very efficiently by Canada during the Winter Olympic Games of Calgary in 1988. The games took place at a number of remote venues. To ensure the 24/7 security of all venues would have required a very large number of personnel that the Canadian police forces simply did not have. A system of deployment was conceived that took into account the following factors: sport events would not take place at the same time in all venues; events would not attract the same interest and hence the same number of spectators; the risks incurred by athletes and officials from different countries were not the same. The role of intelligence was to assess the risks presented by the different events at the various venues. Four different levels of risk were defined as shown in the Table 2.54.

Table 2.54 Four levels of risk defined for a hypothetical event.

Status	Definition	Deployment	Support
Quiet hours	No event.	Video surveillance. Private security guards.	Command center.
Level 1	Event without potential problems or foreseeable risks.	Light deployment.	Reserve available within one hour.
Level 2	Event with potentially limited risks of incidents.	Normal deployment.	Reserve available within 15 minutes.
Level 3	Event with potentially major incident.	Reinforcement of the normal deployment.	Reserve engaged.

Decisions could be made according to the resources available to be deployed. Decisions concerning the deployment of resources were made based on these assessments. A consolidated chart of events was established, indicating for each day and each venue the level of risk. The example below concerns only a specific venue.

Table 2.55 Grid of risks for the hypothetical event.

Hours \ Days	Day 1	Day 2	Day 3	Day 4
Morning	Level 1	Quiet hours	Quiet hours	Level 2
Afternoon	Level 1	Level 2	Level 2	Level 1
Evening	Level 3	Quiet hours	Level 2	Level 3
Night	Quiet hours	Quiet hours	Level 1	Quiet hours

The use of reserves

Keeping a reserve is an old military principle. Reserves increase the commander's influence on the battlefield by bringing in fresh troops to consolidate an advantage or close a gap.

Reserves allow a military or police operational commander to maintain freedom of action and to react quickly to operational developments. Reserves are also a way of addressing the issue of reduction in available personnel.

Case Study: 1940, the French army had no reserves

At the root of many military defeats, one can identify a lack of respect for the reserve principle. In 1940, Winston Churchill, then Prime Minister of Great Britain, was taken aback when he visited the French lines along the German border. The French forces were deployed from the sea to the Swiss border. When Churchill asked General Weygand, the French Commander in Chief, about his reserves, he was told there was none. The subsequent German attack demonstrated how this lack of reserves contributed to the French defeat.

Both of these examples equally apply to police forces, as organizations that usually lack resources and therefore must pursue greater operational efficiency. Two lessons in particular may be drawn. Firstly, the even distribution of force over a territory results in an underlying weakness everywhere; secondly, more flexible deployment of resources compensates in part for any shortage in manpower. These two lessons must be taken into account when redesigning the structures of a police force; they should influence the balance between territorial and mobile forces (e.g. the number of police stations versus mobile patrols) and prompt leaders to examine how mobile reserve force can be established.

Case Study: Reform of border guards in a former Soviet Republic

During the Soviet period, conscripts served in the border guard organization that formed part of the Ministry of Defense. The structures were military in nature, and borders were seen as defense lines. The large number of personnel allowed a strategy where each portion of the border was patrolled. The independence of the former Soviet republic after the fall of the USSR led to reforms in the security system. The border guards were transferred to the Ministry of the Interior and professionalized, meaning manpower was strongly reduced as conscripts were gradually replaced by contracted personnel. As a result, it was no longer possible to control each kilometer of the border. At the same time, border threats had evolved since the end of the cold war, with drugs, human trafficking and terrorism seen as the new security priorities. Therefore, a new strategy had to be conceived that would emphasize border management and the law enforcement aspects of border guard missions rather than the defense aspects. The organization had to move from line defense to area management in an environment where it was no more feasible to keep the entire security force on the border; new operational solutions had to be found such as better surveillance equipment at the border, depth in deployment, flexibility and responsiveness. The only possibility to address the twin challenges of professionalization and new threat was to establish and ensure the availability of mobile forces in each area.

Different methods can be used to establish a reserve capacity.

• A reserve force may be constituted of *specific permanent detachments*, whether incorporated or not into the organization. For instance, the French Gendarmerie comprises the territorial gendarmerie, the mobile gendarmerie and some special branches. The Territorial Gendarmerie carries out public orders and ordinary police functions and is divided into territorial stations (3,600). The Mobile Gendarmerie's principal duty is to maintain public order and personnel may therefore be used as a reserve in the hands of the *Préfets*. The same organization exists within the National Police, with the *Compagnies républicaines de sécurité* (CRS). In Germany, each Land (state) has special units, the *Hundertschaften*, which are characterized by high mobility. At the national level, the *Bundespolizei* (former *Grenzschutz* – border guards) can be required by the Länder to assist their forces in case of large events.

• Reserves can also be formed with *reinforcements sent by other police forces*. In States without a national police force and where the requirements of large events exceed local resources, sophisticated systems of reinforcement can be established. In Switzerland for instance, a gradual system of mobilization has been developed, with positive results (Table 2.56).

• Finally, a police force may also establish its *own reserve* within the organization. A large police force could even create a specific unit for that purpose. Another option would be to prepare the mobilization of police officers through a call system, or summon an *ad hoc* unit formed by police officers from different police stations on a rotating basis (e.g. Friday and Saturday evenings).

Case Study: Efficient use of reserves

The importance of reserves to the freedom of action of an operational leader will depend on his or her ability to use this additional capacity. In 1983, (the situation has changed in recent years) the *Légion de gendarmerie Rhône-Alpe* in Lyon was one of the largest Gendarmerie forces in France. One day, an *escadron* (company) of mobile gendarmes was ordered by the *Préfet* (highest representative of the Government in a department) to move at once to Montpellier in the south of France where some medical students were protesting. The company deployed and came back one day later having only observed the demonstration, which had been peaceful. What was notable, however, was the fact that the commander of the legion could not use his mobile gendarmerie units for actions in his area of responsibility. The suburbs of Lyon are a place where these gendarmes could have been usefully deployed but the colonel could not dispose of them. The unit remained at the *Préfet*'s disposal; only he could provide tasking, with deployment based on political priorities rather than delinquency.

Table 2.56 Constitution of reserves in Switzerland.

Level	Action	Remarks
Police force	If a force cannot handle an event, it may request reinforcements from a higher level.	
Canton	Other police forces may be requested to provide additional manpower.	Municipal authorities must agree (determined by a procedure enshrined in police legislation).
Inter-regional level	Cantons belonging to a region will send reinforcements.	Agreements called "Concordats" exist between cantons in the same region, regulating the request procedure, duties of partners, insurance, costs, etc. The police forces of one Concordat have established a special riot unit whose members are equipped and trained together and which can be mobilized and made available to the canton in need.
Federal level	Reinforcements are sent from every canton according to a previously agreed allocation.	Federal police are tasked with the mobilization of the police officers requested.
Inter-national level	Countries with which Switzerland has concluded agreements may be asked to send reinforcements.	During the G8 of 2003, German *Bundespolizei* sent a few *Hunderschaften* to help the Geneva police.

Case Study: The reserve system in Canada

In order to have reserves in place and to access this additional capacity, a number of options and processes are available in Canada. An agreement operates between the federal and provincial governments (policed by the RCMP), to facilitate the availability of resources. The Provincial Police Service Agreement[33] addresses emergencies within the territory of a province, outside provincial borders but still within the responsibility of provincial authorities, or in an area of federal responsibility.

Where an emergency exists or is likely to exist within an area of provincial responsibility the minister responsible can make a written request to the Provincial RCMP Commanding Officer to redeploy to the extent necessary to maintain law and order, keep the peace and ensure the safety of persons, property and communities.

Where in the opinion of the commissioner of the RCMP that an emergency in an area of provincial responsibility exists or is likely to exist outside the province, he or she – in consultation with provincial ministers and the federal minister – may withdraw up to 10 per cent of the members of the provincial police service, including any necessary equipment, to deal with the emergency. Where in the opinion of the commissioner of the RCMP there is a need to use part of the Provincial Police Service with respect to an emergency in an area of federal responsibility that exists or is likely to exist anywhere in Canada, he or she may temporarily withdraw up to 10 per cent of the Provincial Police Service, including any necessary equipment.

Provisions exist within the agreement for cost recovery and further consultation after a prescribed period of time. This approach has permitted governments and police to operate in the knowledge that if needed, resources are available without having to maintain a large reserve of police at great expense within each jurisdiction.

Partnerships

The idea of enhancing the flexibility of a police force through partnerships rests on the differentiation of tasks to be completed. Certain tasks can be accomplished by non-police officers. In some countries, auxiliaries, specials or supernumeraries, and reserve officers, reinforce police forces. These personnel represent valuable assets in many situations. They can relieve police officers, for example, of control and surveillance tasks. Private security guards may also be hired in such circumstances. In the case of the Calgary Olympic Games, many control and surveillance activities were undertaken by private security officers or by volunteers overseen by police officers.

---

33 Available at *http://www.pssg.gov.bc.ca/police_services/publications/docs/provincial-police-service-agreement.pdf (last accessed on November 22, 2010).*

Equipment

Materials and equipment may enhance the flexibility of a police force. For instance, the new strategy of the former Soviet Republic border guards – centered on less personnel and greater mobility – depends on guards being equipped with enough vehicles and surveillance equipment – sensors and cameras – at the border.

2.25 ISSUES IN THE CHOICE AND USE OF TACTICS

As noted earlier, it is important again to emphasize the importance of *coherence* between a chosen tactic (the way), the problem faced (the end represented by the goals defined) and the resources deployed (the means, considered based on type, quality and quantity). The choice of a specific tactic depends on the *time at one's disposal.* The period needed to prepare and execute a community-policing program cannot be compared to the decision to send a SWAT team to overpower a dangerous perpetrator.

The time element is also the basis for another important consideration: the *degree of preparation* necessary to effectively implement a tactic. This is linked to the complexity of the tactic being considered. To take an example related to the deployment of a combined force composed of reinforcements sent by several different forces, one must take into account the necessary time to (re)train in order to achieve a satisfactory level of professionalism and coherence. A British officer speaking of the tours of a battalion between Ireland and Germany (where it was stationed in the period before 1989) said the professional soldiers needed, at each rotation, weeks to be retrained for the duties they had exercised only six months before. The preparation phase will determine the conditions underpinning the successful use of a tactic. Improvisation may be deadly in such cases as illustrated by the following example.

> **Case Study: The investigator's death**
>
> A dangerous criminal escaped from a Swiss prison in 1983 after taking hostages. His car was stopped and he took refuge with his hostages in an empty building on the outskirts of the city of Lausanne. Subsequently, police surrounded the location and the investigator who had arrested the criminal arrived. Under the impression that he had established a good relationship with the criminal, the investigator thought that he had gained an advantage and would be able to reason with him. On this basis, he advanced uncovered, unarmed and was killed. Police at the time had neither developed tactics to handle such cases, nor specific training for such intervention.

Another important element of the preparation phase concerns the design of *scenarios.* Possible events that could shape, counteract or interfere with the planned action should be identified, and reactive measures developed.

Preparation is also significant in *facilitating* the actions of police – such as, a checklist distributed by the force to an officer, which contains all the elements to be observed when arriving at the location of an incident. One tries to reduce errors and omissions in such cases that might occur due to the stress of a situation. Officers or patrols are expected to take three types of measures:

- Actions: rescue, regulate traffic, protect evidence.
- Organization: define who is in command and who is tasked to disseminate information.
- Information: to whom, what, when, where, why and how.

Experience has shown that what would appear to be evident is not; very frequently, information is not transmitted because officers are concentrating on the work to be done at a site, while the issue of command may also be a problem when several officers arrive together.

The next factor of success is ***training***. Beyond general training, specific training targeted to hone certain actions should be completed when time allows. Training often must be provided at different levels, as shown in Table 2.57.

Table 2.57 Training at different levels.

Who	What	When
Officers	Self defense, offensive defense; Driving; Scene management.	Police academy, refresher courses.
Special units: SWAT, negotiation teams, bomb squad	Specific tactics.	Specialized and refresher courses.
Formation	Crowd control.	Specialized and refresher courses.
Specialized structures	Crisis management.	Specialized and refresher courses.
Staffers	Staff work.	Specialized and refresher courses.
Supervisors	Leadership.	Specialized and refresher courses.
Senior officers	Leadership; Operational and project management.	Specialized and refresher courses.
Personnel involved in specific actions	Necessary skills; Scenarios.	Specific training before actions.

In many situations, the choice of a tactic will depend on the broader *policy or strategic framework* that guides the application. This context will determine the focus placed on the choice of a repressive or preventive tactic. Figure 2.58 places certain policies and tactics on a prevention-repression axis.

The issue of preventive tactics is linked to the *involvement of external actors.* Neighborhood watches are one example, as are security measures taken by private persons or companies to secure their premises. In this context, the growth of private security companies is one of the most important changes to have occurred in the security system in the last years. In many countries today, there are more private security guards than police officers. This has led to a spatial divide between a private and public sphere, the latter being policed by state representatives and the former by security companies. Such a situation may unfortunately lead to the creation of two types of security beneficiaries.

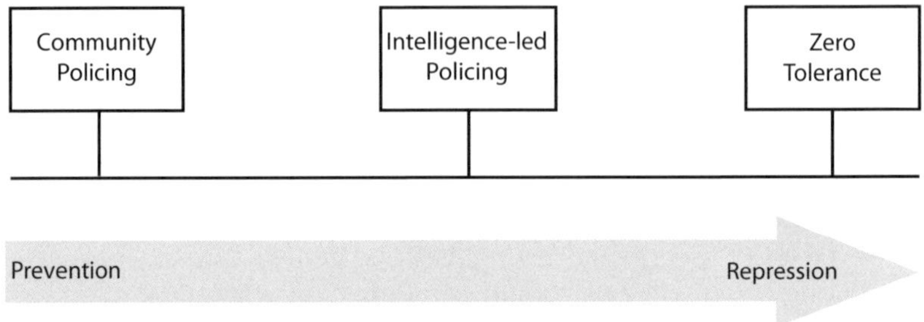

Figure 2.58[34] Prevention/repression axis.

Legal factors will also delimit the possible use of a tactic or its use in a given context. For instance, the use of weapons is restricted by several conditions. The manner in which the law is applied is as well an important consideration: that is, strict application versus flexibility in situations where police are granted some discretionary power to decide – notably in public-order oriented operational settings (see the Annex) – not to intervene when subsequent actions could produce more harm than good.

Information management is a principal component in the determination of a tactic and subsequent execution. Though the importance of the links between information, decision, execution and preparation for the future have already been explained, it should be re-emphasized that the chief of operation and the specialized members of his staff must clearly request the data to be searched for in their areas of responsibility.

Table 2.59 includes some examples of information required for decision-making in specific areas.

[34] Pioneered by former New York City Police Commissioner, William Bratton, the Compstat Process is a strategic crime-control technique, centered around four crime-reduction principles: accurate and timely intelligence, effective tactics, rapid deployment of personnel and resources and relentless follow-up and assessment. Fundamentally, the Compstat process is based on the principle that by controlling serious crime, police are better positioned to maintain order and solve other problems in the promotion of safety.
(Extract from Shane, Jon M. Compstat Process. *FBI Law Enforcement Bulletin*, April 2004, pp. 12-20.)

Table 2.59 Information required for decision making.

Area	Types of information
Definition of a community policing strategy.	Environmental data.
Definition of a problem oriented policy.	Data specifically related to the problem to address.
Crowd control.	Before the event: advanced information on protestors, intentions, equipments; During the event: on line information.
Ongoing operation.	Information for monitoring.
Opinion and expectations of members of the force and of citizens.	Opinion surveys.

A final important factor for success is an *assessment of the potential negative effects* of the tactics chosen, in order to minimize them if possible. For instance, the increase of controls on a street that may be caused by a crackdown operation in an area may also provoke some frustrations amongst the population. Information and explanation may be necessary to reduce the possible anger of the citizens.

2.26 SUMMARY AND KEY POINTS

- Tactics are ways of achieving strategic goals efficiently using available means.
- Four principles have to be taken into consideration by the decision-maker when choosing tactics: flexibility, concentration, adequate allocation of resources, and diversity of action.
- The search for greater flexibility in the use of available resources is a constant preoccupation of an operational leader.
- Reserves, a sound balance between generalists and specialists, delegation of powers, management by operational objectives are all key factors for enhancing this flexibility.
- The main factors of success in the choice and execution of tactics are coherence between ends-ways-means, good preparation, adequate training, and appropriate management of information.
- When implementing a given tactic, different factors have to be taken into consideration: the situation, the time available, the aim of prevention versus repression, legal aspects, and the potential negative effects.

ANNEX 2.1
POLICE TACTICS, TOOLS AND INSTRUMENTS

In this Annex, a selection of tactics will be briefly described in considering the level at which they can be used.

Table 2.60 Policies and strategies in tactics.

Tactic	Definition
Problem-oriented policing (POP)	The emphasis in POP is on directing attention to the broad range of problems the community expects the police to handle. POP recognizes that the ultimate goal of police is not simply to enforce the law, but to deal with problems effectively – ideally, by preventing them from occurring. POP encourages consideration of a wide range of alternatives, in addition to criminal law.[35] Operations conducted within a POP framework can be repressive or preventive in nature. The main elements of success of a POP strategy are a good understanding of the problem, a mix of adequate solutions addressing different aspects – thus, the use of a well-structured method is important – and cooperation with the different actors involved. Communication and consultation to obtain information is necessary to identify the problem, fix priorities and measure results (in terms of achieving a reduction of the problem).
Community policing (CP)	CP is a collaborative effort between law enforcement and the community, with the aim of identifying problems of crime and disorder that involve all elements of the community, in the search for solutions. Skogan speaks of a triumvirate of citizen involvement, problem solving and decentralization.[36] CP is chiefly a preventive strategy and its main goals are to increase the trust of the public in the police and to satisfy the needs and expectations of citizens. Key factors of success are a trustful relationship between the public and the police and the adaptation of police structures to this form of policing. A data-driven strategy is also an important element of success. Results are measured according to the level of satisfaction of the community.
Zero tolerance policing (ZT)	ZT is a repressive approach that emphasizes the absence of any leniency or exception in the enforcement of the law and regulations. This approach states that if minor incivilities are ignored, an atmosphere of lawlessness and social decay will appear, creating an environment where higher forms of crime will develop. Key factors of success are the strict application of the law and an adequate justice system to make ZT effective. Results are measured according to the diminishing of the targeted incivilities and the level of satisfaction of citizens.

35 Herman Goldstein cited in Jerry Ratcliffe, *Intelligence-led Policing*, Willan, Portland, Oregon, 2008, p. 70.

36 Wesley G. Skogan, "The promise of community policing" in: David Weisburd and Anthony A. Braga (eds), *Policing Innovation: Contrasting Perspectives*, Cambridge University Press, New York, 2006, pp. 27-43.

Table 2.60 Policies and strategies in tactics (cont.).

Tactic	Definition
Intelligence-led policing (ILP)	ILP is the application of criminal intelligence and crime analysis as an objective decision-making tool in order to facilitate crime reduction and prevention through effective policing strategies and external partnership projects drawn from an evidential base.[37]

Table 2.61 Operational tactics.

Tactic	Definition
Specific campaigns	Repressive or preventive operations conducted in given areas for a certain duration in order to address specific problems such as driving while intoxicated and drug trafficking or use at schools.
Crackdowns (CD)	CD is repressive and dissuasive operations where the police concentrate substantial resources, for a limited time, on targets (areas, persons) clearly identified as causing serious problems. CD must be thoroughly prepared – an objective-driven collection of information is a prerequisite – and sufficient resources must be deployed. The key factors of success depend on the preparation, the concentration and the dynamism of the police (the police presence in the area is not sufficient, police officers must act: control, arrest). The surprise effect is important for the first operations but the publicity given to a CD approach, when conducted over a certain time, can amplify its effects and be dissuasive. CD by definition and practice involves limited action meant to have immediate impact on the problem and to gather further information. CD can hence be considered as specific measures within a Problem-oriented policy.
Crowd control	One of the tasks of police is to maintain – or if necessary, restore – order. Tactics for controlling a crowd and preventing the outbreak of disorder have been developed for that purpose. They may be preventive or repressive. The image of police has often been severely damaged by the methods involved in accomplishing this mission. The development of community policing in the USA can be seen as a reaction to the brutality of some police interventions during the urban riots of the 1960s and against the often peaceful anti-Vietnam war demonstrators. Therefore, new tactics for maintaining and restoring order had to be developed and sections of police forces specially trained and equipped for these missions. (In some countries, special troops are tasked with the maintenance of order such as the CRS, the gendarmes mobiles in France or the Territorial Support Group of the Metropolitan Police.)

[37] Jerry H. Ratcliffe, Intelligence-Led Policing, *op. cit.*

Table 2.61 Operational tactics (cont.).

Tactic	Definition
Crowd control	Tactics used can be preventive (barrages, access controls, presence of riot control police) or repressive (encirclement, pincer movement, kettling also known as containment or corralling, sweeps, and charges.) Armament and material are adapted to the situation and gradually deployed: barriers, tear gas, rubber bullets, plastic bullets, beanbag rounds. The influence of two legal principles has an effect on the use of tactics. The *principle of legality* states that police have to intervene with appropriate means as soon as the law or a regulation has been violated; for instance, if demonstrators begin to encroach on the forbidden area or throw projectiles at officers. This situation assists in establishing a clear line between what is accepted and what is not, but at the same time, may inflame situations and provoke conflicts. The legal *principle of discretionary* power allows police the flexibility to employ a measured response to a demonstration and demonstrators. This approach takes into account the activities, mood of the demonstrators and public safety of all considered in order to avoid inciting the situation. Police tactics have been adapted to meet ever-changing challenges as demonstrations become more frequent, more violent and better organized. Small groups of violent demonstrators, decentralized and operating with agility and will, on the model of urban guerillas, harass police. They have also mastered the manipulation of the media for their own purposes. Hence, police have had to become more flexible and replace or at least complement traditional static deployments with snatch squad tactics and other similar actions. Key factors of success have been improved intelligence gathering and infiltration of violent demonstrator groups. The UK police deploy, for instance, Forward Intelligence Teams to gather intelligence on the ground and in some circumstances to disrupt violent activists and deter anti-social behavior. Cameras, camcorders and audio recorders are used to conduct overt surveillance of the public. Decentralization of decision-making is also a principal element in the management of demonstrations where violent demonstrators operate in a similarly decentralized manner (although key decisions – such as the use of plastic bullets – may remain in the hands of the chief of operation).

Table 2.62 Tactical Tools

Tactic	Definition
Police vehicle skills and tactics	Chiefly encompassing: • Police safe-driving techniques for all officers as well as advance driving skills for pursuits, surveillance, and VIP driving for those engaged in such activities. • The methods for stopping and searching vehicles.
Random patrols[38]	The Kansas City Preventive Patrol Experiment[39] has shown that the increase of police patrol activity in an area does not necessarily reduce crime or alter citizen attitudes concerning the police or fear of crime. Police today tend to increase the efficiency of patrol activities by assigning targets based on the results of intelligence analysis or providing officers with useful crime information concerning specific areas of patrol.
High visibility patrol	These mainly preventive patrols aim at increasing patrol visibility and can take many forms – for example, walking beats, vehicle patrols, officers on bicycles – in concentrating on specific areas where crime may be most impacted by increased police visibility. The identification of targeted areas through intelligence analysis is a key element of success as well as the operational management of deployments: the composition, frequency, and location of patrols should be changed frequently.
Low visibility patrol	The aim of this type of repressive tool is to deter street crime by increasing the rate of arrest of offenders. Undercover agents and plainclothes patrols may be used, while the element of surprise and information are key elements of success.
Directed deterrent patrol	This form of patrol has a preventive purpose, with deployments based on prior detailed criminal analysis for a specific area. Hence, the time and location of a patrol will be fixed according to the information provided.
Target-oriented patrol	These patrols are directed toward specific individuals, locations, or events, combining elements of high-visibility, low-visibility, and directed deterrent patrols. Information and thorough preparation are key elements of success.
Split-force patrol	This is a technique in which part of a patrol executes preventive measures and the remaining members respond to service calls.

38 Charles D. Hale "Police Patrol, Operations and Management", Pearson Prentice Hall, Upper Saddle River, 2004.

39 George L. Kelling, Tony Pate, Duane Dieckman, Charles E. Brown "The Kansas City Preventive Patrol Experiment: A Summary Report", Police Foundation, 1974, http://www.policefoundation.org/pdf/kcppe.pdf (last access September 21, 2010).

INVESTIGATION TACTICS

Investigation tactics can be understood in a broad sense as the various methods applied to systematically address key questions in the aftermath of an offence, linked to six interrelated areas (Table 2.63).

Table 2.63 A guide to investigation tactics.

Questions	Tactics/tools
WHAT has happened?	• Interviewing techniques; • Interrogation methods; • Physical evidence; • Profiling; • Use of records; • Physical surveillance: fixed and mobile; • Informants; • Undercover agents; • Technical surveillance: wire-tapping, oral or written communication; • Investigation (deductive and inductive processes).
WHERE has it happened?	
WHEN did it happen?	
HOW did it happen?	
WHO is involved and in what way?	
WHY did it happen?	

Investigation has also been defined as "the collection of information and evidence for identifying, apprehending, and convicting suspected offenders."[40] On the other hand, Ralph F. Turner of Michigan University stated that, "a criminal investigation is the reconstruction of a past event."[41] Investigations can be divided into three phases: first, the identification of perpetrators; second, ascertaining the location of the suspected offenders; and third, preparation of the case to be submitted to court. In the initial phase, patrol officers arriving first on the scene play a key role in the success of the investigation. The police investigator in the second phase plays an equally primary role in the identification and arrest of perpetrators and will take charge in the final phase (in collaboration with prosecutors).

[40] E. Ostrum, R.B. Parks, and G. Whitaker, *Patterns of Metropolitan Policing*, Ballinger, Cambridge, MA, 1978, p. 131.

[41] J. W. Osterburg, R. H. Ward, *Criminal Investigation*, Anderson Publishing Co., Cincinnati, OH, USA, 1992.

SPECIFIC MECHANISMS

Table 2.64 describes operational instruments that have developed over time that use specific linked tactics.

Table 2.64 Use of linked tactics in operations.

SWAT TEAMS	SWAT teams are special groups of police officers selected and extensively trained to deal with unusually dangerous or violent situations, such as hostage-taking or heavily armed individuals. SWAT members are highly trained with enhanced weapons capabilities and may be organized into permanent teams – for instance the French GIGN (*Groupe d'intervention de la gendarmerie nationale*), the German GSG9 (Grenzschutzgruppe 9) – or work generally as regular police officers while remaining on call.
Negotiation teams	Negotiation teams are comprised of specially trained officers called in when hostages have been taken or a person is threatening harm to themselves or others. The negotiation teams' role is to establish contact and dialogue with the suspect to defuse the situation or bring it to resolution without the use of force.
Sharp shooters (or snipers)	Sharp shooters are generally deployed with SWAT in hostage scenarios alongside negotiators and an assault team. They are trained to shoot only as a last resort when there is a direct threat to life.
Bomb squads	Bomb squads are responsible for responding to incidents involving explosives. Bomb technicians are highly trained with specialized equipment to conduct preventive checks and safely deal with explosives and incendiary devices.
Crime scene management	Crime scene management is an investigative support, and concerns all measures undertaken when: • Arriving at a scene; • Preparing preliminary documentation; • Processing a scene; • Completing and recording the crime scene investigation.
Use of informants and undercover agents	An informant is someone who provides privileged information about a person or organization to a law enforcement agency without the consent of that person or organization. An undercover agent is a police officer disguising his or her identity, or using an assumed identity, for the purpose of gaining the trust of an individual or organization to obtain information or evidence. The management of informants and undercover agents raises a number of specific problems related to legality, supervision and control.
Technical surveillance	Technical surveillance encompasses all the technical possibilities that a police force can use to gather information or control individuals or locations.

FOURTH PART OF THE OPERATIONAL DECISION-MAKING CHAIN: *EVALUATION*

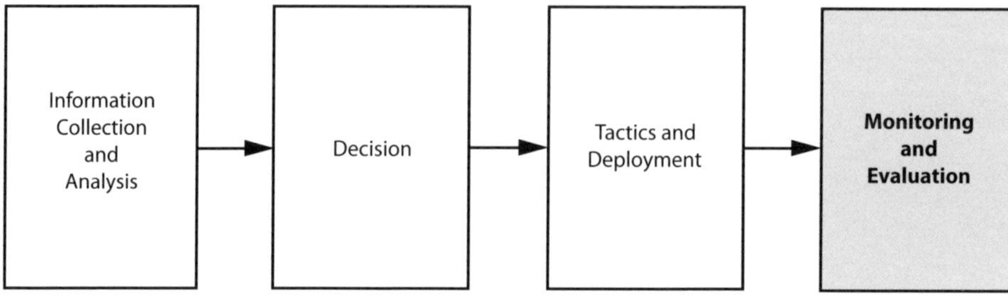

Figure 2.65 Evaluation phase of the decision making chain.

Objectives

This Section will discuss different issues related to evaluation (ends, ways and means) with a focus on lessons learned and their use for correcting identified weaknesses.

2.27 DEFINITIONS

2.27.1 Control, assessment, evaluation, monitoring, oversight

Each of these terms refers to the regular observation and recording of project, program or operation-based activities undertaken within an organization, to ensure these remain in line with broader strategic goals. Subsequent reporting enables the information gathered to support corrective measures at different levels (including ongoing activities, strategic reviews and organizational arrangements) and/or assist in the identification of lessons learned and best practices. With this in mind, the four general aims of evaluation activities are:

- to detect differences and changes to a situation, a behavior defined as standards, a planning, or an objective;
- to provide a basis for taking corrective measures;
- to assess results; and
- to enrich the process of organizational learning.

The term *control* will be used here to refer specifically to the process of ensuring the overall coherence of procedures, budgets, schedules and results in certain organizational settings. In addition, *control* in this context alludes to the concept of regulation, the process of bringing to uniformity various mechanisms towards general or specific objectives. Similarly, *assessment* refers to evaluation and *monitoring* as well as the tracking of a project or operation in the implementation phase. *Oversight* will be reserved here to describe the supervision exercised by political and legal authorities (executive, legislative, or judiciary) over public organizations, or by boards of directors and shareholders on public companies.

2.27.2 System of information

Control can be viewed as an element of the system of information; that is, a means to assess how an organization fulfils its stated mission in addressing the risks, requirements and the expectations of a particular operating environment in the process of delivering services or products. Figure 2.66 illustrates the links between intelligence (the role of which is to identify risks, needs and expectations), control procedures that monitor the level of productivity (comprised of effectiveness and efficiency) of an organization in delivering services, and communication (including internal communication aimed at enhancing strategic awareness, and external communication centered on increasing levels of public trust).

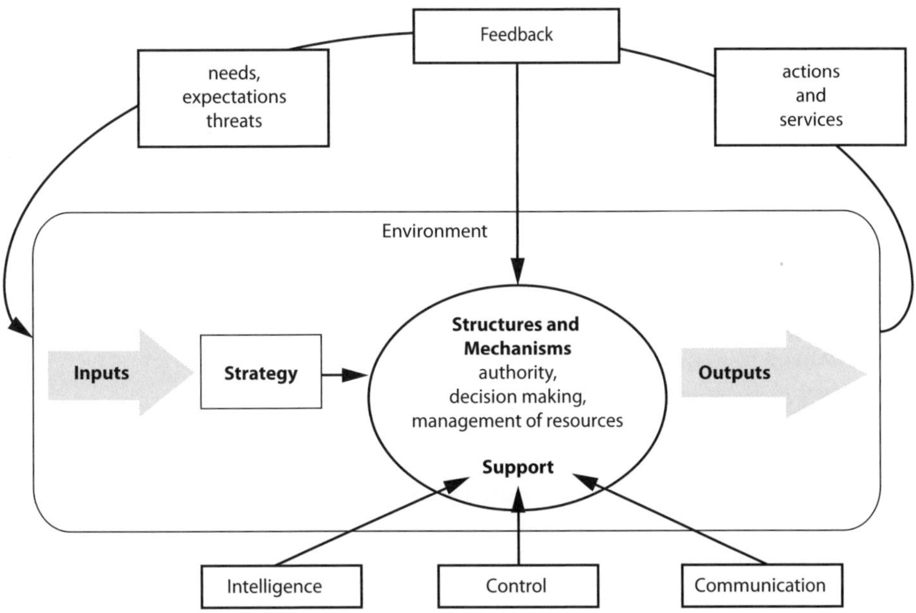

Figure 2.66 View of the system of information.

2.38 SYSTEM OF CONTROL AND OVERSIGHT

2.28.1 Types and levels of control

Turning to the specific context of police organizations, Figure 2.65 represents the different types and levels of control and oversight. One can clearly distinguish external from internal controls. The first category refers to oversight exercised directly by the executive, legislature and judiciary, and indirectly by the media and civil society.

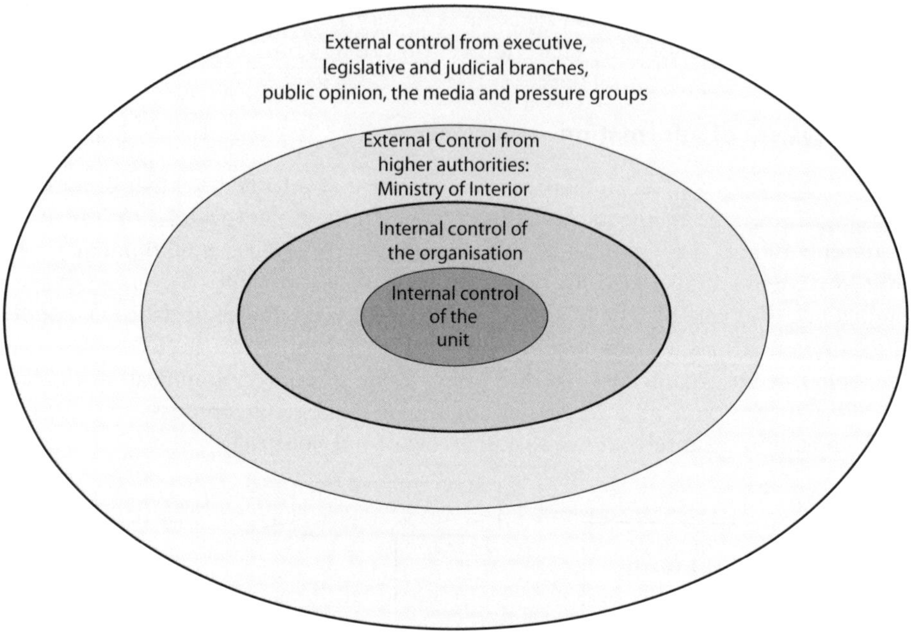

Figure 2.67 Levels of control.

The system of control applied within police organizations must be able to meet the information requirements of external stakeholders (for instance the Ministry of Interior, but also the media, and the public). The system will naturally prioritize, however, the supply of data and information needed for monitoring police activities.

Building upon the four aims of evaluation activities outlined on p. 117, one can identify five key *purposes* for systems of police control and oversight:

- Effectiveness: Have objectives been achieved?
- Efficiency: Are outputs (results) commensurate with the inputs (resources) invested or deployed?
- Legitimacy: Are the measures taken legal and proportionate?
- Accountability: Have relevant actors (police officers, managers, senior officers) been held accountable for their own actions towards their organization and others?
- Equity: Are all subjects of police activity treated equally and ethically?

2.28.2 Tools of control

Table 2.68 outlines some of the main tools used to support control and evaluation activities.

This chapter will further develop two of these tools: audit and performance evaluations.

Table 2.68 The main tools used in evaluation.

Tool	Purpose	Method
Inspections	*To verify* actions, execution of orders and tasks.	Controls by the chain of command; Complaints against officers;
Controls	*To verify* the consistency of: Procedures; Budgets; Deadlines; Results.	Controls by the chain of command; Reports; Check lists.
Assessments	*To assess* management. *To assess* productivity. *To assess* performance. *To assess* programs. *To evaluate* personnel. *To measure* public trust.	Audit (financial, managerial); Process evaluation; Programme evaluation; Impact evaluation; Personnel evaluation; Statistics; Surveys.
Expectations and needs	*To identify needs.* *To compare* services catered to expectations and needs.	Surveys; MBWA (managementt by walking around); Interviews of officers as to their effectiveness in carrying out assignments.
Culture	*To influence* behaviour according to internal norms and values.	Training; Peer pressure; Role models.

Audit[42]

The aim of an audit is to sketch a picture of an organization at a given moment, in order to identify and evaluate relevant strengths and weaknesses in light of opportunities and threats faced. An audit can also be limited to highlighting deficiencies in procedures, structures and other organizational aspects. As a result, one of the key factors of success of an audit surrounds the definition of terms of reference. Certain aspects of an audit, especially framing questions and the internal and external dimensions are developed further in the Annex 2.2.

Performance evaluations[43]

Great importance should be attached to performance evaluations, which should usually address three issues:

- How a police officer accomplishes his or her mission (the identification of weaknesses that should be corrected).

---

42 The annex to this chapter presents a more detailed audit system.

43 Charles R. Swanson, Leonard Territo, Robert W. Taylor, *Police Administration*, Pearson Education Inc., Upper Saddle River, 2005, p. 394.

- The selection of personnel for promotion (officers should be evaluated according to the demands of a new position rather than their performance to date; pinpointing strengths that can be developed and weaknesses that should be corrected, are also important elements of the evaluation).
- Identification at the force level of weaknesses reflected across individual evaluations that need to be addressed via modified or new training programs.

A best practice approach to performance evaluations should ensure:

- A focus on performance standards (linked to job descriptions).
- Ratings based on clearly defined criteria.
- The communication of relevant standards is understood and constitutes the subject of evaluation.
- The incumbent chief presents the results of the evaluation to the officer in question (this offers an opportunity for both to discuss performance, work conditions and potential areas of improvement, as well as to clarify the future expectations of both parties).

Evaluations are generally done by completing the adequate forms that evaluate performance, but in certain cases, notably as part of the selection process for higher ranks, written evaluations are often complemented by other tools, such as assessment centers (where role-play exercises, written exercises and oral presentations enable evaluation of candidates "in situations.")

Control as a management tool

Control can be understood as a management tool; the systems involved assist managers in monitoring the functioning of their organization, the progress of projects and delegated actions and the identification of weak spots. Control measures also support processes of reform in highlighting areas for improvement.

Despite having their origins in the business world, the following three management tools can also be easily applied in a police environment.

The Management Cockpit

The Management Cockpit, developed by the neurosurgeon Patrick Georges, is a simple and proven method for improving the performance of teams, departments and entire companies. This improvement is achieved because employees are motivated by focusing on what is most important. Improved performance is consolidated by measuring activities, because what gets measured gets done. A Management Cockpit displays information to a manager the way an airplane cockpit presents information to the pilot. Just like a cockpit, the Management Cockpit provides the instruments a manager needs to run a business unit. In a Management Cockpit room, all the major engines for making decisions and monitoring implementation are accessible: knowledge, decision-makers and means of action.

In Patrick Georges' words, "It is just like flying an airplane. If you want to take off, you must first review the checklist. All the indicators relating to the task at hand must be green. For example if you wish to decrease inventory to reduce your unliquidated assets, the Logical View will tell you that the delivery timetable is between two to five days. This is a lot considering that our competitors deliver within 24 hours."[44]

The Balanced Scorecard

The Balanced Scorecard (BSC) is a, "strategic performance management tool; a semi-standard structured report supported by proven design methods and automation tools that can be used by managers to keep track of the execution of activities by staff within their control and monitor the consequences arising from these actions. The core characteristic of the Balanced Scorecard and its derivatives is the presentation of a mixture of financial and non-financial measures each compared to a "target" value within a single concise report. When measures are reviewed, it is possible to determine whether current performance meets expectations. The idea behind this is that by alerting managers to areas where performance deviates from expectations, they can be encouraged to focus their attention on these areas, and as a result trigger improved performance within the part of the organization they lead. The balanced scorecard is a management system (not only a measurement system) that enables organizations to clarify their vision and strategy and translate them into action. It provides feedback around both the internal business processes and external outcomes in order to continuously improve strategic performance and results. When fully deployed, the balanced scorecard transforms strategic planning from an academic exercise into the nerve center of an enterprise."[45]

Case Study: The Balanced Scorecard at the RCMP "G" Division

The Royal Canadian Mounted Police adopted the BSC in 2002. The "G" Division (the RCMP is divided into 15 divisions, and the letters define the area of responsibility of the Division) set up a working group to create an accountability framework, define priorities and prepare the implementation of the BSC. The new concept was intended to achieve three objectives:

1. Allow the organization to set real performance expectations and keep all staff on track in meeting those expectations.
2. Ensure the entire organization was working towards common and compatible goals.
3. Provide a means of measuring how close the organization came to meeting goals and expectations.

---

44 Patrick Georges, "How Management Cockpits Facilitate Managerial Work" and "The Management Cockpit, A First Experience", downloaded from http://www.patrick-georges.net/en/download_22.php

45 http://www.balancedscorecard.org/BSCResources/AbouttheBalancedScorecard/tabid/55/Default.aspx (last access November 22, 2010)

The commanding officer of "G" division defined what he thought the benefits of the new system to be. It would permit management to measure progress and help to build credibility by tracking, and then communicating efforts and success. It would also help members of the division to see for themselves how he or she was doing. The commanding officer stressed the importance of the BSC as a tool for obtaining resources by convincing senior management and donors of the needs. Moving from a strategic to tactical application, the commanding officer stressed the beneficial use of the BSC as a foundation for guiding day-to-day management, human resource deployment and budgeting. The creation of a strategically focused organization would be made possible with the aid of this tool, shown in Table 2.69.

Table 2.69 BSC approach of the RCMP.

The division was currently unable to	In the future the division aimed at
Provide basic, quality performance information to the right people at the right time.	Providing key results information to assess ongoing performance.
Make strategic decisions using performance information.	Making strategic decisions using performance information.
Ensure external or internal accountability.	Comparing performance against that of others in order to gauge the level of relative organizational achievement. Achieving greater accountability to Government, partners, the public and all levels within the organization.
Demonstrate its ability to deliver value for money.	Demonstrating its ability to deliver value for the money invested in the organization.
Fully justify existing or new resources.	Demonstrating clear value for new resources and funds being requested.

The rating system

Based on the rating attributed by banks to companies according to their financial security, a similar system has been developed for police by a consultant company, Team Consult of Geneva in cooperation with ASSEKURATA, a German Insurance Company. The idea is to assess police forces based on a similar set of criteria, in order to identify the areas in which they outperform or underperform their contemporaries. Such benchmarking is intended to support the development of recommendations for organizational reforms, as well as help in establishing contacts between forces to facilitate the exchange of effective solutions.

2.28.3 Data for control

The collection of relevant data must be conducted based on the objectives set for control. This should involve the definition of indicators, the tasking of individuals responsible for collecting information, and clarification of the frequency and form of data presentation, as well as the necessary audience.

The dangers of control

One must, take into account the risks involved in the implementation of a system of control. Firstly, the costs of such a system are high in human as well as in financial terms. The system must therefore address real needs in an economic way and not generate a set of figures not directly relevant to the issues faced. Secondly, controls can negatively influence members of the organization who may resent the measures involved as infringements upon their area of responsibility that limit their initiative. Thirdly, the potential for incoherence between the system of control and the stated goals of the organization can risk the spread of cynicism or disillusionment amongst personnel, undermining the credibility of management and discrediting control measures. In such contexts, controls can send the wrong signals.

Case Study: The wrong signals

A neighborhood watch project was launched in Switzerland, with a number of information sessions organized. The meetings were very successful and many citizens offered to volunteer. Unfortunately, one evening, the traffic department had not been made aware of the meeting held in a certain location and had chosen the area in question as a site for the installation of radars. Thus, a certain number of people who had volunteered for the project were caught by the radar and fined. It is difficult to criticize such individuals for thinking that the police, like Janus, had two faces.

A fourth category of risks concerns potential errors in the focus of control[46], which are frequently encountered in a policing context where, for understandable reasons, compliance with administrative and legal procedures is essential. Yet, a focus on compliance may lead to a neglect of performance. Thus, a delicate balance must be established in order to develop an efficient system of control.

---

46 Meaning, what is controlled is not relevant to overall strategy. For instance, the focus is placed on the respect of the rules and procedures and the results are not accounted for.

Case Study: Use of indicators

The use of indicators may also produce negative and unintended effects. A police chief decided bonuses would be linked to the number of arrests or tickets his police officers conducted on the street. This led to a situation where the officers under his command multiplied their arrests/tickets without reason, raising public resentment through their zeal, which considered their actions as police harassment.

Some years ago in France, the Minister of Interior, decided to get rid of community policing actions in some difficult suburbs because he did not believe it was the task of police officers to play football with local youth. The minister introduced a "results culture": police stations (commissariats) had to achieve greater efficiency. The number of arrests, crime statistics and other indicators became the criteria by which officers were evaluated, promoted, displaced or sidelined. Initially, this policy seemed to work, but soon it was realized the figures were misleading: complaints were no longer registered in some cases, statistics were embellished and results were frequently not achieved according to the priorities given but in areas where it was easy to do so. Winston Churchill's words, "I only trust figures that I have forged myself" were particularly applicable to the climate created. In such cases, the focus shifts quickly from quality, to quantity and ease of achievement.

2.28.4 Specific problems relating to control within police organizations

A system of control, like structure, follows strategy. The system must be designed in a way that supports implementation. Thus, it is critical that the subjects of control are clearly identified and the optimal tools chosen. Focusing in particular on police organizations, one must take into account the special character of the activity and structures involved.

On a whole, police forces remain hierarchical despite attempts in certain instances to rid the rank pyramid and empower officers (especially in their relationships with the public through the influence of community policing). The necessary empowerment of police forces to establish valuable contacts is not facilitated by an often times closed culture and distrust of the outside world (reflected in an "us against them" mentality). At the same time, contemporary police forces cannot disregard the scrutiny exercised by the media and public, which ensures a level of accountability for any errors, abuses or violations. In this context, compliance with written procedures is important, though one must always remember that in all decisions, and measures taken by police officers the human factor remains the chief element – hence the importance of monitoring and evaluating this aspect of police activity. As a result, management must anticipate the main areas of potential problems and develop a system of control accordingly.

Some of these "risky" areas of activity include:

- SWAT teams and special units, which tend to develop as strong subcultures within the force and in their external relations. Their aura and status as highly trained specialists may lead members to think that rules and regulations must not interfere with efficiency, which is their highest aim. Managers should seek to reduce such risks through careful squad selection to eliminate "Rambo" types, through ethics training, regular assessment, rotation of personnel, close supervision, and thorough debriefing in the wake of completed operations.
- Informants and undercover agents: the risks involved depend on their autonomy and powers given; their activities may infringe on particular legalities. In many countries, the use of these agents is therefore subject to the authorization of a prosecutor.
- Traffic patrols because they are vulnerable to bribes in many countries.
- Drug squads, which are often required to deal with the temptations involved in handling large sums of money and quantities of drugs.
- Interviews must have controls that focus on their legitimacy and their compliance with the rules; adequate training is one way to avoid such problems.
- Weapons use, which usually constitutes a delicate issue and hence requires multi-faceted forms of training (shooting, rules of engagement, adaptation of training taking into account lessons learned).
- Incentives, gifts and other rewards (for instance, free coffee at gas stations, discounts on meals) which, even if of a minor nature, can lead police officers to offer reciprocal benefits.
- Off-duty activities, though control in this domain is delicate as one could easily infringe on the private sphere of personnel. The development of an effective supervisory relationship represents the preferred method for anticipating potential problems.

Clear rules, ethical codes and mentoring, as well as tailor-made monitoring mechanisms, are necessary to manage risks.

2.29 USE OF ASSESSMENTS

Any evaluative data collected must be analyzed in order to highlight discrepancies between initial targets and subsequent achievements, and serve as the basis for improvements. Data can be derived from the different documents established from interviews, inspections, or other techniques as described in the Table 2.69. Analysis however should not always rest on figures; it must also take into account the on-the-job experiences of police officers. This is why debriefing procedures should be systematically conducted after all large, complex or targeted operations. The objective of such debriefing should be linked to the identification of operational strengths and weaknesses, and to clearly define what has worked and what has not.

The following example, which is interesting to consider in this context, comes from the military.

Case study: Military knowledge management[45]

The U.S. Army's Opposing Force (OPFOR) is a 2,500 member brigade whose job is to help prepare soldiers for combat. Each month, the brigade simulates the role of a hostile army, or insurgents, paramilitary units or terrorists against a new brigade of more than 4,000 soldiers in the California desert.

OPFOR has developed the After-Action Review (AAR), a method for extracting lessons from one event or project and applying them to others. The OPFOR commander's operational order commences the cycle. The order contains four parts: task to be accomplished, purpose (importance of the task in the given context), commander's intent (how the task will be done) and the end state (what the desired result is). On this basis, each unit will establish a clear understanding of what it has to do and how. It will also develop IF/THEN scenarios.

The AAR addresses four questions to identify what worked and what did not: What were our intended results? What were our actual results? What caused our results? And what will we maintain or improve? The point here is primarily to correct *thinking* than correct *things* because learning will affect the unit's ability to plan, adapt and succeed in future battles.

The US Army uses formal knowledge systems to capture and disseminate lessons to large, dispersed audiences, for instance with teaching and training by the National Training Center; e-mail and Internet; and through officers and NCOs seconded to other units.

The AAR has become an important tool in many companies, but is not always executed with the same rigor as by OPFOR, with the risk that the AAR is turned into a *pro forma* wrap-up. The failures of the process stem from misconceptions about the nature of the AAR, that it is a report and postmortem, used mainly for blaming rather than for improving collaboration and organizational units' learning. The problem is not only to extract lessons from experience, but more to learn how to use such lessons to adapt and act.

A business version of the AAR, customized for the corporate environment, could require teams to answer four questions before embarking on important action: What are our intended results and measures? What challenges can we anticipate? What have we or others learned from similar situations? What will make us successful this time? The responses to these questions can serve to align the team's objectives and set the stage for an effective AAR meeting following the action.

Ultimately, one can identify four key success factors in such a process:

- The AAR process must start at the beginning of the activity.
- Lessons must first and foremost benefit the team that extracts them.
- Lessons must link explicitly to future actions.
- Leaders must hold everyone, especially themselves, accountable for learning.

---

47 Marylin Darling, Charles Parry and Joseph Moore, "Learning in the Thick of it", *Harvard Business Review*, July-August 2005.

The following example is taken in a police environment:

Case study: The RCMP System

Within the Royal Canadian Mounted Police, there are currently four levels to the accountability framework. The internal audit unit provides senior management assurance regarding the soundness of controls, risk management practices, information for decision-making, as well as the relevance, effectiveness and efficiency of programs across the RCMP. A dedicated group of professionals conducts these objective assessments and the results are presented to the Departmental Audit Committee (DAC). This approach allows management decisions to be guided by the identification of issues, opportunities for improvement and high-level controls that could mitigate risk.

Regional reviews and program reviews are conducted throughout the organization at the request of the heads of business lines or regional senior management. These reviews are an important part of any ongoing program management cycle and are intended to provide performance information regarding regional or program operations.

Management Reviews (MR) are an independent assessment of unit management and assess the adequacy and effectiveness of a unit's management and investigative performance. By providing regional, divisional and business-line managers with risk-based assessments of administrative and operational practices, these reviews form a fundamental part of the RCMP's governance process.

Unit Level Quality Assurance (ULQA) is an ongoing, internal, self-monitoring tool. Annually, the RCMP units select unit-level activities for a quality assurance review through formal risk analysis and prioritization to ensure monitoring effects are applied in areas of risk. There are five phases to the ULQA process (RCMP, 2007b):

1. Identification of activities.
2. Risk management and prioritization of reviews.
3. Physical review and observation.
4. Corrective action.
5. Follow-up.

Management reviews should focus on tasks that are fundamental to a unit, achieving objectives through good management as well as the application of the operations model and the adequacy and effectiveness of quality assurance processes. The steps include:

- Selecting units for review based on a risk assessment.
- Conducting management review and report on results and recommendations using trained reviewers.
- Monitoring corrective action and promote sound practices.

Taken together, the RCMP four- level approach provides a range of tools that allows the organization to address both local operational needs and corporate strategic goals.

The results of assessments based on an analysis of data and debriefings may have implications at three different organizational levels, as shown in Figure 2.70.

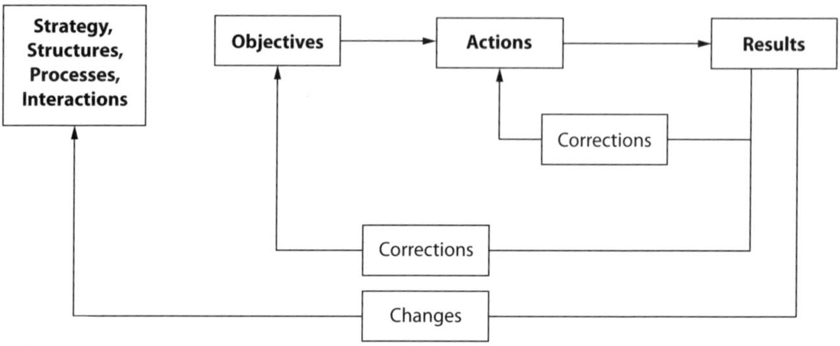

Figure 2.70 Assessments and their implications.

Corrective measures concern ongoing actions for short-term measures. The aim of such measures is to redress what is going wrong in specific areas of an operation or project. At a higher level, the objectives of an operation or project may have to be reviewed as operational progress or broader shifts decrease their continued relevancy. Finally, at a strategic level, reforms may be necessary in order to ensure an organization (a system) adapts to the challenges it faces, as highlighted by analytical assessments of a broader scope.

Case Study: G8 Meeting, 2003

Even though the results of the G8 meeting held in Evian in 2003 were good, debriefings revealed shortcomings in the Swiss security system. The reports prepared after the event detailed problems at three different levels of intervention. Each of these constituted an area where reforms were proposed; the higher the level, the more far reaching the reforms suggested:

- At the level of the police forces deployed, amendments to procedures and tactics were proposed.
- At the level of collaboration between cantons, the establishment of coordination structures and rules of cooperation were proposed.
- At the national level, necessary changes to the entire security system and related security policy were highlighted.

While corrective measures have since been taken at the first two levels, improvements have still to be accomplished at the national level.

2.29.1 Specific difficulties in evaluating police

One should be aware of the specific difficulties in evaluating police activity. Primarily, the difficulties are linked to the problem of defining the criteria of success: is it the number of arrests, of case clearances, or of the number of offences? Is a police force that clears cases at a higher rate better than another force where the number of offences within its jurisdiction is lower? There is no correct answer, as many of the conditions of success do not depend on a specific police force, but instead on the features of the environment in which it works. Secondly, difficulties arise because of the resistance of police officers and police organizations as a whole, to being assessed. Thirdly, it is often difficult to evaluate the "real" causes of an improvement or deterioration in the level of crime; what is the influence of police operations as compared to other factors (for example, environmental changes in the economy, demography, etc.). As a result, researchers have tried to find other ways of evaluating the success of police activity, focusing on what has worked and what has not (Table 2.71).

Table 2.71 WHAT WORKS, WHAT DOES NOT WORK[48]

Type of tactic	What works
Focus on most severe crimes and criminals.	The focus on recidivist criminals seems interesting, but a focus on drug trafficking is less efficient. Intensive inquiries focused on criminals with prior offenses succeed in increasing the degree of imprisonment. Operations tailored to address specific problems, notably illegal arms possession on hot spots or drugs trafficking in buildings work.
Patrol deployment at specific points.	According to the researches done in USA, this strategy is efficient in resolving problems at the local level.
Focus on repeated victims.	Research in the UK has shown that this measure can significantly affect certain crime types.
Collaboration between different authorities.	According to research in the UK, this could be a very useful working model for police.
Campaign against impaired driving.	Campaigns against impaired driving involving intensified controls[46] and arrests have proven successful.
Adapting operations to problems encountered.	This approach was tested on a small scale, but a formal evaluation of its impact on criminality should be made on a larger scale.

48 This table was inspired by a chart presented by Jérôme Ferret in «Evaluer la police de proximité?», *Les cahiers de la sécurité intérieure*, Paris: IHESI, 3e trimestre, 2003, p. 91. Ferret's chart is an adapted version of a chart by, Peter Jordan "Effective Strategies for Reducing Crime" in Peter Goldblatt and Chris Lewis (eds), *Reducing Offending: An Assessment of Research Evidence on Ways of Dealing with Offending Behaviour*, Home Office Research Study 187, Home Office, London, 1998, pp. 65-66. Elements were also taken from Etienne Blais and Maurice Cusson "Les évaluations de l'efficacité des interventions policières: résultats des synthèses systématiques", in: *Traité de Sécurité Intérieure*, *op. cit.*, pp. 115-129.

Table 2.71 WHAT WORKS, WHAT DOES NOT WORK (cont.).

	What doesn't work
Random patrols.	Radom patrols are inefficient.
Community policing.	Generally inefficient, unless the objective is to increase police legitimacy in the eyes of the local population. For example, surveillance of neighborhoods and citizen assemblies brought together by the police, two measures commonly associated with community policing, do not have an impact on criminality. It should be noted, however, that much research has been undertaken on the efficiency of community policing and the results have often differed.
Informal contacts with children.	Generally inefficient.
Improving solving rates.	No real influence on the level of crime.
Quick response to citizen calls.	A quick response to citizen calls is not necessarily followed by an increase in arrests.
	Differentiated results/undetermined
Increased police deployment.	A simple increase in the number of police officers on a territory does not have an impact on criminality. However, the almost total absence of police officers, for example due to a strike, is almost always accompanied by a wave of crimes.
Increased utilization of arrest power.	Efficient in certain cases of spousal assault; counter-productive for minors.
Rapid responses to emergency calls.	Diverse results; a study in the US judges this measure as inefficient while another study in the UK considers it to lead to only marginal improvement in the resolution of burglary incidents.
Routine patrols.	Patrols, during which officers police the illegal possession of firearms, seem promising, but further evidence is needed.

The US National Research Council Panel (NRCP)[49] reviewed research concerning police effectiveness and measured the effectiveness of different strategies. The NRCP's report concluded that the "standard or traditional model" of policing demonstrated little or no evidence of effectiveness; the Community Policing Model showed a weak to moderate effectiveness concerning community relationships; and Problem-oriented Policing demonstrated strong evidence of effectiveness concerning crime hot spots. The research concluded that focused policing and a combination of different approaches seemed to produce the best results. To reaffirm, this research constitutes one finding amongst many, and conclusions may differ in other studies.

---

49 Wesley Skogan and Kathleen Frydl (eds), *Fairness and Effectiveness in Policing*, National Research Council, The National Academies Press, Washington DC, 2003, pp. 248-249.

2.29.2 Factors of success for a system of control

Factors of success for a system of control can be grouped in four categories, as shown in Figure 2.72.

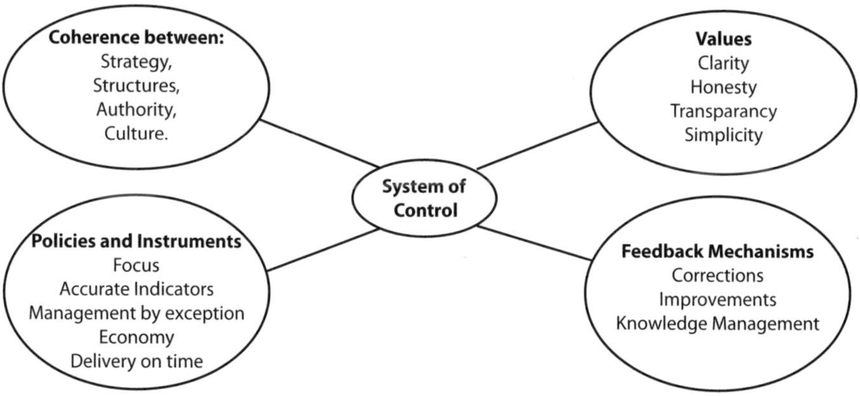

Figure 2.72 Factors of success for a system of control.

2.30 SUMMARY

- Control (evaluation, assessment, monitoring and oversight activities) is part of the system of information.
- Police forces are overseen by political and judicial authorities, as well as by the media and public.
- There are five key objectives related to the control of police, and different tools can be used for these purposes.
- The Cockpit Management and Balanced Scorecard systems represent powerful management tools.
- For a system of control to be effective, it must be coherent with the strategy *and the structures* of the organization.
- A system of control must remain economic, and therefore concentrate on exceptions. Certain areas and activities of police forces require particular attention within a system of control.
- The results of evaluation and assessment activities can be utilized at different levels; these results can form the basis for corrective measures to be taken during an operation in process; they can lead to re-formulated objectives and they can also launch strategic organizational-level thinking.
- Evaluation of police performance is difficult, though one promising approach is to consider what works and what does not.

ANNEX 2.2
AUDIT

Table 2.73 Six fundamental questions to be answered.

Question	Purpose of the assessment	Example: central dispatch center
Why is the activity undertaken?	To define the purpose of the activity and to assess if it should be modified.	Analysis of the mission of the CDC. Should it be the responsibility of the CDC to filter calls/take the first measures, or should these tasks be done by others?
What are the activities undertaken to achieve the purpose?	To establish the chain of activities in order to assess if the process can be streamlined.	In relation to the mission as defined above.
How are the activities undertaken?	To assess if the activities could be completed in a more efficient way (simplification of procedures, technological help).	Are there computer-aided decision programs that could be applicable?
Who is undertaking the activity?	To assess if the right people are in the right place/if the profile of those who undertake the activities are appropriate (over- or under qualified).	What kinds of profiles are needed? Are all the necessary skills present? Should the personnel of the CDC receive specific training? Can civilians be engaged for this activity?
Where are the activities undertaken?	To examine if measures of decentralization or centralization could be necessary.	Are there several CDC in the force? Could their number be reduced, modernized, expanded?
When are the activities undertaken?	To assess if the frequency of the activity is appropriate.	Can the staff and or the shifts be reduced at certain times of the day/night?

INTERNAL ASSESSMENT

The *overall objective* is to assess if there is a clear strategy addressing the security challenges arising from the police environment and in line with the population's expectations and needs, and to examine if the existing resources are both adequate, and effectively and efficiently deployed. Table 2.74 lists of areas form the basis of the audit.

Table 2.74 Areas that form the basis of an audit.

Issue to be audited	Key points of the audit
Strategy	Is there a clear strategy and plan of action? Does it correspond to relevant challenges and capabilities? Is the process for establishing the strategy adequate? Is the strategy known by all staff and personnel? Is a process in place for regular evaluation and updating of the strategy?
Missions and tasks	Are the missions clear? Does an adequate legal framework exist? Are police undertaking tasks that could/should be outsourced?
Values	Are the values clearly stated, taught and known? Does a values chart, code of deontology or of ethics exist? Are issues of gender, youth, and minorities adequately taken into account? Are existing means of compliance with rules respected? Is there an effective system to handle complaints against police?
Operational policies	What are the operational policies in different areas: community policing, patrols and response to calls, criminal, crisis management, public order? Do adequate regulations and standard operating procedures exist? Are debriefing procedures in place and lessons learned managed (databank, consequences for training, changes of procedures, etc)
Organizational structures	What are the structures of the police force? Do the structures align with the strategy and operational policies? Territorial coverage? Response time? Mechanism for fixing priorities? Job descriptions?
Systems of authority and delegation of power	Does the system of authority match the policy and structures?
Resources (human, material, financial)	Are resources adequate in relation to the problems and tasks? How are resources managed? Human resources: recruiting, training, promotions, evaluation of required skills and competences, evaluation of personnel, gender policy. Equipment: adequacy? Procurement methods.
Learning system	How does the force learn from experiences and use these constructively for future planning?
Administrative mechanisms and procedures	What are they? Are they adequate?

Table 2.74 Areas that form the basis of an audit (cont.).

Issue to be audited	Key points of the audit
Intelligence and knowledge management	Intelligence cycle. Exchange of information with other agencies. Legal framework. Links with the decision-making process. Adequacy of material.
Internal and external communication policy	Challenges. Policy. Resources.
Control systems (management, finances, behavior, procedures, values)	Statistics: are they relevant and not overly time consuming?

EXTERNAL ASSESSMENT

The *overall objective* is to assess the nature of relations with external partners and how these help or hinder organizational performance (Table 2.75).

Table 2.75 External assessments.

Issue to be audited	Key points of the audit
Environment and analytical methods.	How is the environment analyzed? Are risks and opportunities evaluated? Are the implications for the police force identified?
Image of the police in the eyes of authorities and public at large.	What measures are in place to survey and promote the image of police?
Relations with official authorities: executive, legislature, judiciary, other state agencies.	What are the links between these bodies? Do clear mechanisms exist for the exchange of information? Are roles clearly defined and respected?
Relations with the public.	Does a policy of community policing exist? Are regular surveys conducted? What cooperation arrangements are in place with local authorities? Is there a system for fixing priorities? How are complaints handled?
Relations with other police forces (national, international).	How do the relations function? How is information exchanged? How are policing activities coordinated?
Relations with other security actors: the army, border guards, private security companies, intelligence agencies.	How do they function? How is information exchanged? How are activities coordinated?

Chapter 3

Structures

Chapter 2 is devoted to examining the different phases of the decision-making process. Decisions are not made in a vacuum; structures, managerial organization, styles of leadership, to name a few of the environmental factors of the decision process – influence not only the way decisions are made but also how they are implemented. They represent key factors of success or failures. It is in light of this that these elements will be addressed in present and following chapters.

Objectives

This chapter will focus on the organizational structures that influence decision-making and implementation. Links between strategy and structures will be highlighted and models of administrative and operational intelligence structures will be presented and discussed.

3.1 STRATEGY AND STRUCTURES

In military terms, *strategy* is the art of marshaling and deploying resources to achieve the aims determined by political authorities. In a business context, strategy is the policy adopted by an enterprise to ensure long-term profitability.[1] In the ends, ways, and means model presented in Chapter I, strategy is the art of combining the ways and the means to achieve the ends. Figure 3.1 describes the general process of strategy development.

---

1 For that purpose, answers to the following questions will have to be given by the management: In which business are we in, what should be our business, and how do we get there? This means to agree on the mission (the WHY), the vision (the WHAT), and the plan of action (the HOW).

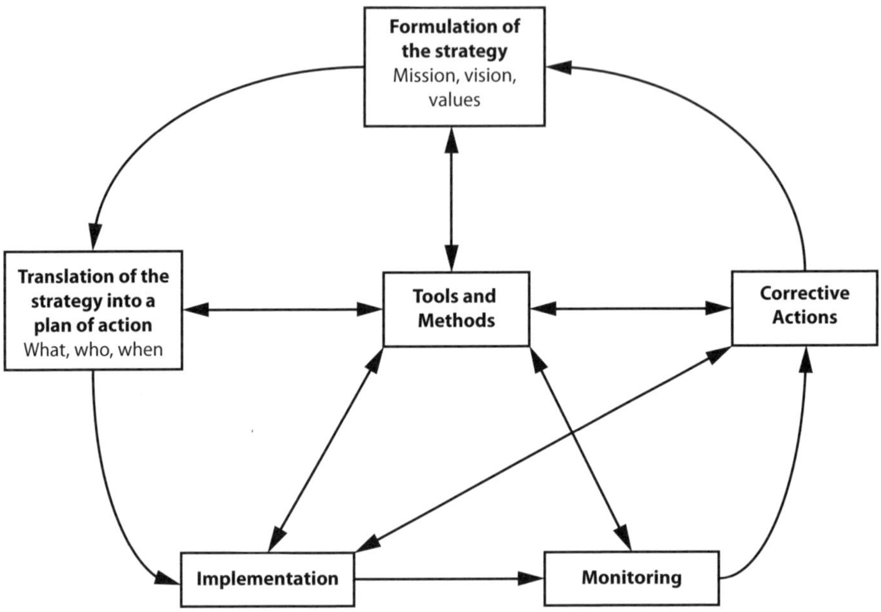

Figure 3.1 General process of strategy development.

Case Study: The strategic planning process of the RCMP

The Royal Canadian Mounted Police has developed an Integrated Planning Cycle structured on the following process (Figure 3.2):

- In the spring of each year, an assessment of past performance is completed, in addition to a preview of the coming year and up to a five-year horizon. A directional statement and business plans for the year to come are established.
- In the fall of each year, mid-year performance is assessed and adjustments are implemented for the next six months;
- In the late winter of each year, the business plan is reviewed for the coming year and the budget confirmed for an April start.

The planning cycle, in order to fulfill the strategic goal of "Safe Homes, Safe Communities" focuses on:

- The strategic priorities fixed by the RCMP (in the areas of organized crime, youth, international police services, etc).
- The strategic objectives linked to each function: for instance, prevention and education, intelligence, investigation, enforcement, and protection.
- The management strategies aiming at, for example, improving communication, human resource capacities, planning and implementing performance management.

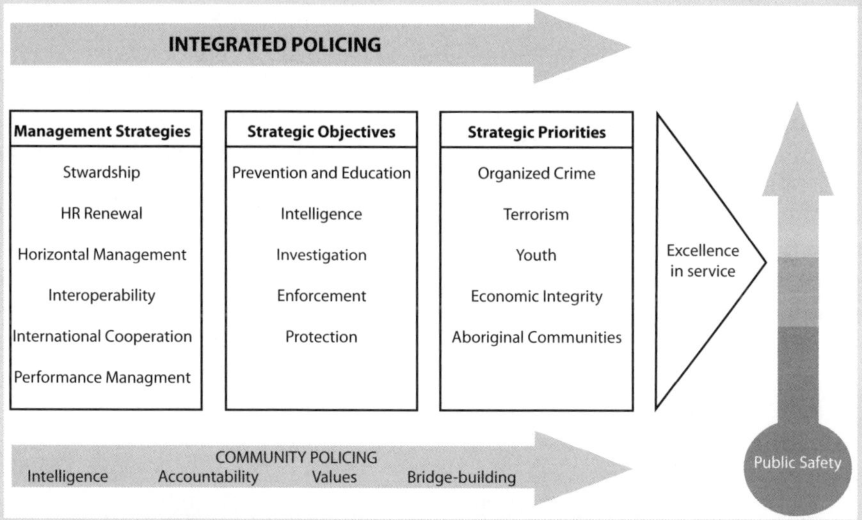

Figure 3.2 Strategic framework.

Source: http://www.rcmp-grc.gc.ca/about-ausujet/strat-eng.htm (last access October 26th 2010)

The planning cascade can be represented as shown in Table 3.3:

Table 3.3 Planning Cascade of the RCMP.

Strategic Planning	• Federal Government priorities; • Portfolio priorities; • Environmental scan; • Operations issues.
Business Line	• Strategic plan; • Business line priorities; • Regional/operational/tactical input; • Federal government initiatives.
Regional Business Planning	• Strategic plan; • Business line plans; • Contract partner's plans/expectations.
Operational and Tactical Planning	• Regional plans; • Multidisciplinary teams; • Mandated responsibilities (contracts, MOUs); • Targets of opportunity.

A decision, at each level – strategic, operational and tactical – must be analyzed, made and implemented within given or *ad hoc* structures. As previously stated, an important principle states that structure follows strategy. This principle cannot always be respected as in many cases strategy development must take into account existing structures. Thus, one should rather speak of the interactions between strategy and structure. For example, two countries undertook deep reforms of their police system in recent years. Each sought to achieve more efficiency and rationalize costs. The political structures of the countries were very different, however, and the solutions implemented had to consider structural factors. Austria had (until 2004) four independent security forces that all reported to the Ministry of Interior: gendarmerie, criminal police, security police (*Sicherheitswache*) and border guards. These four entities were merged into one police force spread across nine regions (Figure 3.4). The strategy adopted by Austria, a unified police force decentralized on a regional basis, corresponds to the country's long history of centralized administration and respects contemporary territorial organization.

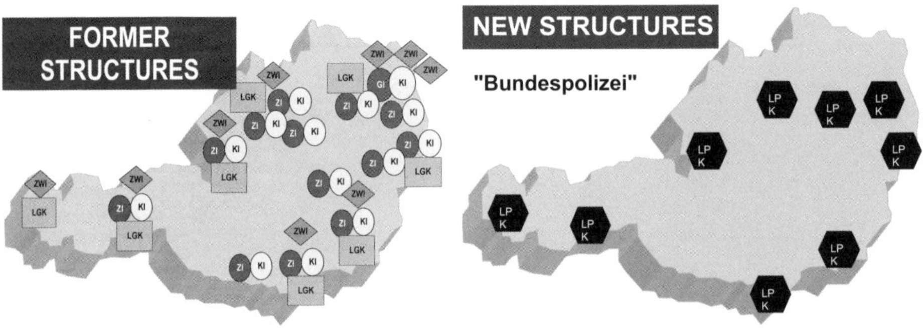

Figure 3.4 Austria and its structures.

In the second case, Belgium, the power of the *bourgmestres* (mayors), who were strongly opposed to losing their grip over local police forces, compelled the government to perpetuate, and even increase, *the presence* and power of the city police. Today, the federal police coexist with 196 police zones reporting to local mayors. A complex system of coordination has been established in order to reduce the difficulties created by the existence of 27 regions regrouping each several zones. Each region has two directors, the director of the judiciary police and one director whose task is to ensure cooperation between his or her zone, the other zones, and the federal police. An important factor linked to the overall coordination of the different elements lies in the *Plan National de Sécurité* (Security National Plan) which is established every four years. The plan is based on the "Image of criminality" which takes into account traffic, incivilities, crime and terrorism, and considers the impact of these phenomena and the threats they represent in taking into account the vulnerabilities of society. Ten topics are then identified as key subjects for the government to consider when establishing police priorities. Each zone then develops its own four-year security plan; the zone is not obliged to replicate all the governmental priorities but must adapt them to its own situation. If necessary, the zone can add its own priorities. The regions must prepare annual plans that deal with the concrete measures to implement in order to achieve the set priorities.

Case Study: Mutual Links Between Strategy and Structure

In the 1990s, the 1000 members of the Police of the Canton of Vaud (Switzerland) were located on three sites (notwithstanding the distribution of decentralized police stations), which housed, respectively, the gendarmerie, the criminal police, staff and support services. The gendarmerie were located a few kilometers from the criminal police, which were housed in a building close to the premises in which the headquarters and support services were established. This dispersed arrangement created many difficulties; in particular, the relationship between uniformed and plain-clothes police officers suffered, with the gendarmerie and criminal police developing operational autonomy with negative consequences for the effective functioning of the force and running costs. The chief later succeeded in securing funding to regroup the whole force in one area. The new constructions were designed to reduce or suppress the problems identified. Building entrances, indoor traffic and recreation areas were conceived in such a way that contacts between members of the force were fostered; a new distribution of offices allowed the merging of redundant administrative services. Positive impact was soon observed: sharing and exchange of information improved, administrative costs diminished and a certain number of officers could be returned to operational tasks. These results had been made possible because a clear strategy had first been defined and then implemented with a regard to structure.

3.2 GENERAL STRUCTURES

How should one approach the design of organizational structures? The process can be described as follows:

- Activities are analyzed.
- Activities are regrouped into building blocks (Drucker groups them according to their contribution: activities producing results; support activities; maintenance activities; top management activities).[2]
- Decision processes and relationships are identified.
- Coordination mechanisms are established.
- The structure is put in place.

Organizational design focuses on two levels of differentiation: the level of authority (vertical) and the level of activities (horizontal); activities can then be grouped by type of clients or services, by geographical areas or by processes.

The following section introduces three basic types of organizational structures and considers their main advantages and disadvantages.

---

2 Peter Drucker, *Management, Tasks, Responsibilities, Practices*, Geneva: Management Editions (Europe) Dirmas S.A., 1974, 517-599.

3.2.1 The functional structure

The functional structure (Figure 3.5) is built along the main building blocks of the police force. Its principal advantages are its simplicity, clear definition of tasks, responsibilities and specialization. The functional structure however has several disadvantages, including a tendency to develop centrifugal forces and resistance to change; coordination may also be difficult. The focus may rest too heavily on the management of divisional problems at the expense of general strategy.

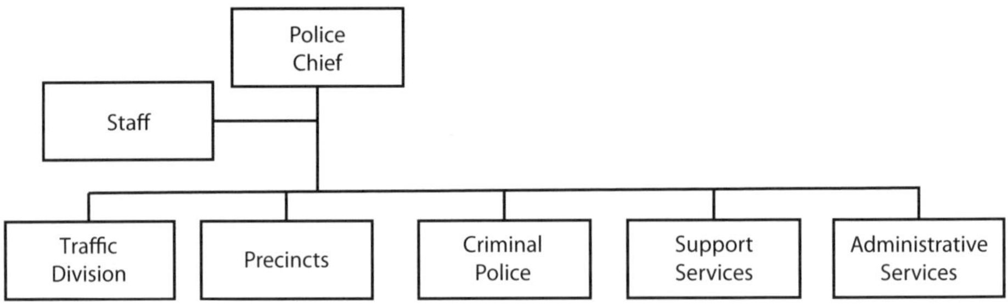

Figure 3.5 Functional structure.

3.2.2 The divisional structure

The divisional structure (Figure 3.6) is organized across divisions by products, processes or geography (a police force in a federal state for instance). The essence of a matrix structure lies in the assignment of members to functional areas (such as patrol, detective, or support services), to specific projects (e.g. task forces and crime-specific programs).[3] The principal advantages are to be found in the fact that the chief may concentrate on general strategy and management skills, which are developed at several levels. Correspondingly, among the main disadvantages, one can point to the risk of conflict between divisions and the complexity of coordination.

3.2.3 The matrix

The model with is referred to as the matrix combines a functional and divisional structure (Figure 3.7). One of the key advantages of the matrix is linked to the increased responsibility placed upon managers, leading to the development of their management skills. However, the risk of conflict between divisions is real, and issues of coordination and control are complex, as specialists would be required to report to two chiefs: a functional chief and an administrative manager.

---

[3] Charles R. Swanson, Leonard Territo, Robert W. Taylor, *Police Administration*, Pearson Education Inc., Upper Saddle River, 2005, 247.

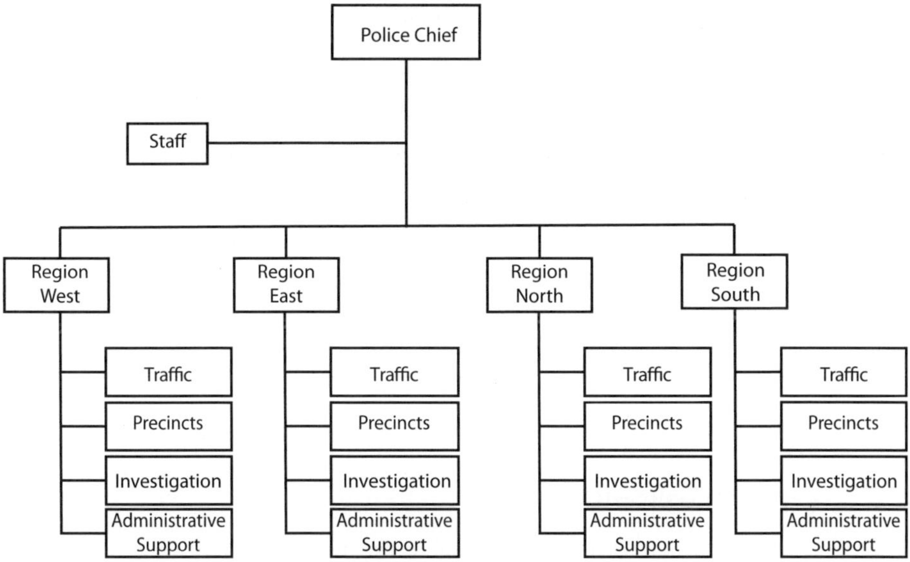

Figure 3.6 Divisional structure.

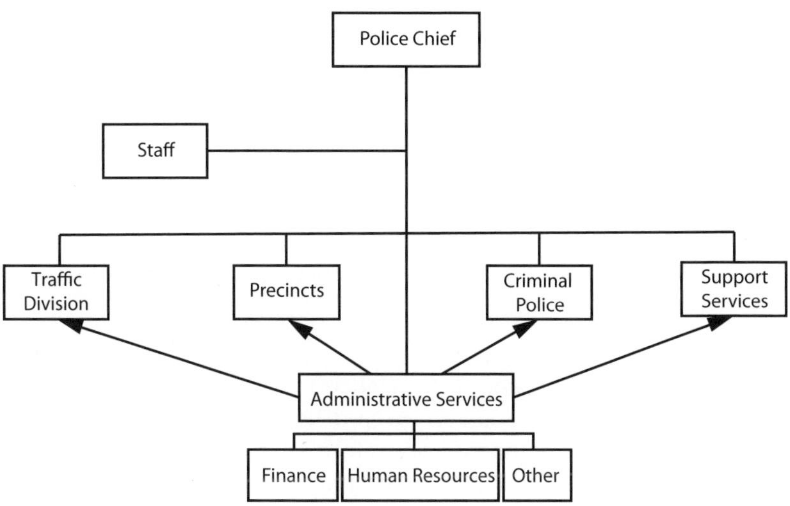

Figure 3.7 Matrix.

3.2.4 New types of organizations

Against the backdrop of social evolution and globalization, businesses have been compelled to adapt in order to gain more flexibility and efficiency in regard to organizational forms and managerial processes. These constraints are reflected in the development of task forces, of decentralization within organizations and of networked organizations. Such changes have also influenced police forces.

3.3 INTELLIGENCE UNIT STRUCTURES

The efficiency of intelligence collection and analysis depends on many factors, including adequate organizational structures. For this reason, it is important to devote special attention to the issue of establishing an intelligence unit within a police organization. In doing so, one must differentiate between organizational and operational intelligence units. The first is a permanent unit integrated into the police organization chart; the second forms part of an *ad hoc* structure (generally prepared and trained) mobilized in specific situations. The key elements of such a unit include function, integration in the general organization, structures (administrative and operational) and personnel requirements. A model administrative order outlining important aspects to be dealt with in the establishment of an intelligence unit is provided in an Annex to this chapter.

3.3.1 Levels of intelligence units

Intelligence processes can be categorized according to the different scope and application of data collected; organizations must therefore develop appropriate procedures to ensure effective coordination between these levels. Problems can become very complex according to the situation. The management of information is, for instance, more difficult in Belgian police forces than in Austria, due to the number of different autonomous entities. Intelligence is generally managed at three levels:

- The *national level*, where the intelligence function is strategic in focus, analyzing threats, trends and issues concerning the whole country (and establishing links with foreign partners).
- The *operational level*, where the intelligence function is focused on a geographic region.
- The *tactical level*, where the intelligence function is focused on data collected locally.

Case Study: Criminal Intelligence in Switzerland

Switzerland is a federalist country covered by 26 police forces (cantonal). In addition, two metropolitan police forces have investigative powers (Lausanne and Zurich). A federal police force also has specific jurisdiction related to international police cooperation, internal coordination, and conduct investigations, mainly related to organized crime.

In this complicated system, intelligence has been roughly organized in the following way (Figure 3.8):

- Strategic analysis is mainly carried out at the level of the confederation. Some cantonal police forces also practice strategic analysis.
- Operational analysis is undertaken at a regional and local (canton) level. Since 1994, a regional analysis center (CICOP) collates and cross-references information on high volume crimes from the whole region in

order to deal with mobile crimes. The center disseminates a series of targeted intelligence products to investigators, chiefs of operations, and other partners (such as border guards). Similar analysis centers have been developed in the period since, covering the whole country (Zentra, RLZ, PIZO, see Figure 3.8). Each of these centers has recently adopted a similar doctrine for analyzing and classifying repetitive crimes.

- Tactical analysis is organized disparately by each police force, from the federal police to the relevant metropolitan police, in support of investigations. However, a common training program for the whole country was recently implemented.

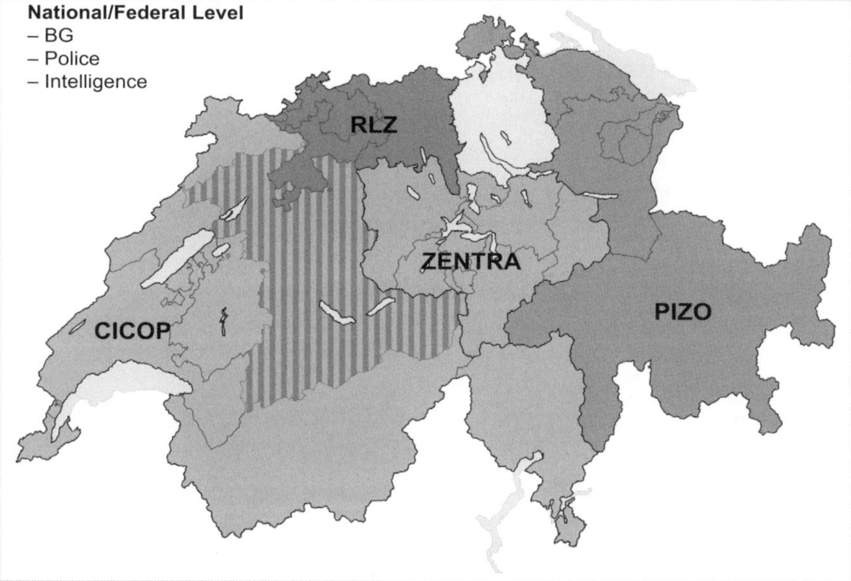

Figure 3.8 Levels of analysis in Switzerland: the smallest entity represents the canton (internal borders).

This system has demonstrated great advantages in the understanding of crime problems that cross jurisdictions. However, coherent functioning of the different levels – from local to central – both from an analysis and decision-making viewpoint, is still far from being achieved. For example, most police still lack awareness of the intelligence-led style of policing.

In contrast with the police system, border guards are the responsibility of the Swiss federation. The national headquarters (BG in the figure above) carries out strategic intelligence, while each regional command has an intelligence unit collecting data provided by the different border guard stations. Each of these stations has a border guard officer tasked with the collection of information and its transmission to the regional intelligence unit, where data is consolidated and transferred to the national headquarters.

3.3.2 Key elements of an intelligence unit

Though the requirements for establishing a permanent intelligence unit in an organization are applicable for all types of intelligence units, missions, tasks and resources will depend on the specific level considered.

3.3.3 Basic prescriptions

- A *policy* states the purpose of intelligence; it is the basis for the development of a *strategy*.
- The *mission* of the organizational intelligence unit is formulated: for instance, to collect data requested by the management, to register and analyze them in order to assist decision-makers in their understanding and assessment of situations, development of options, decision, and monitoring of implementation. An operational unit would usually perform the same activities, though applied to certain specific situations.
- The *structures* of the unit take into account its main *functions*: collection, analysis, developing intelligence products, and dissemination.
- *Job descriptions* state the contributions and tasks of individual unit collaborators.
- The *mechanisms of coordination* and interaction are described (within the unit, with other parts of the organization, and with outside partners); the *outputs* (intelligence products) are defined (form, frequency, etc), as well as *the audience and method of delivery*.
- The mechanisms of *control and feedback* allow for performance review; they also gauge the satisfaction of those using intelligence products.
- Support elements are addressed, especially the *selection and training* of personnel.
- Rules of *security* are established.

3.3.4 Integration and cooperation

One of the key factors of success for every intelligence system lies in establishing an efficient flow of information. The importance of an integrated intelligence approach cannot be overly emphasized, as a large diversity of sources (inside and outside the organization) will nourish and enrich the analysis. The 5W + H questions, outlined in Chapter 2 can be used in identifying sources, methods, and recipients.

A successful intelligence system also rests on the development of an efficient network of sources and relations, not only outside the organization but also internally. Unfortunately, it is typically considered that information sharing within an organization is hindered by competition and jealousy between units, inadequate organizational processes and structures, poor communication and feedback processes.

The sharing of information on a regional basis, when different police agencies share a single territory, is a component of success supporting efficient cooperation.

The leading agency should be responsible for the development and management of the system, analyzing the information provided by other agencies (the sources), disseminating the products of analysis, and if possible, coordinating the subsequent actions. Fusion centers can also be established, with a goal of coordinating resources across overlapping local jurisdictional boundaries.

Case Study: Regional Information Sharing System (RISS)
Many small agencies cannot fund full-time intelligence units. The RISS has been created in the US to assist intelligence sharing. RISS is a regional network comprised of six multistate centers designed to operate on a regional basis and provide support to its members in information sharing resources, analytical services, training and technical assistance. RISS serves over 8,100 law enforcement and criminal justice agencies representing close to a million sworn officers working in local, state and federal agencies in the USA, Australia, Canada, and England.

The development of intelligence cooperation is also one of the main features of efforts to enhance police cooperation at the international level. Some of the most important cooperative institutions are:

- *Interpol* was founded in 1923 and has 186 members worldwide. Interpol maintains databases on known criminals, lost and stolen passports, fingerprints and DNA profiles. Information is shared via a network of national central bureaus.
- *Europol* gathers and analyzes intelligence on organized crime. Member states send police officers to its headquarters in the Hague, which shares information directly with national police forces as well as with a hub of Europol crime analysts.[4]
- *Frontex* is an EU agency based in Warsaw tasked to coordinate operational cooperation between member states in the field of border security (its activities are intelligence driven).
- *Bilateral agreements* on establishing border contact points such as the *Centre de cooperation policière et douanière* between France and Switzerland where Swiss and French police officers, gendarmes and custom officers are housed in a common building at the Geneva airport in order to share information and coordinate action.
- *Liaison officers* and *police attachés* play an important role in bilateral and multilateral police cooperation.

4 Fred Schreier, "Fighting the Pre-eminent Threats with Intelligence-led-operations," Occasional Paper No 16, Geneva: Geneva Centre for the Democratic Control of Armed Forces, 2009.

Case Study: Cooperation between military and police in Peace Support Operations (PSO)

If armies and police forces have often developed efficient cooperation in responding jointly to major disasters, it is in the context of PSOs that a need for improved collaboration has arisen. Ideally, PSOs should progressively evolve from a significant international military deployment, which hands over to an international civilian police component, which in turn trains and then gradually passes authority to the domestic police service. Ideally, this is a process of increasing appropriation by local stakeholders and the implementation of democratic control of the police.

However, reality is different, and the phases described frequently overlap. Therefore, it is necessary to clearly establish the relevant cooperation framework. Three generic 'cases' can be identified:

- *In place of*: This occurs when the military phase is achieved but a regular police force has not been deployed, or not entirely. As a result, a vacuum develops that inhibits the ability of the operation to maintain order. The army is not equipped and trained for this type of mission and whenever possible paramilitary police forces (*gendarmerie, carabinieri*) must be deployed.
- *In support of*: This occurs when a police force has been deployed and the army provides logistical and operational support for actions conducted by the police.
- In cooperation with: This occurs when police forces and armies cooperate in joint operations.

Experiences to date from PSOs have highlighted many difficulties relating to military-police cooperation, including mutual prejudices, cultural differences, different problem-solving approaches, a lack of common operational procedures, and inadequate information sharing.

Measures should be taken in order to improve coordinated planning prior to the deployment of PSOs to train senior personnel in advance; to establish a common language procedures and mechanisms of coordination; to develop information exchange; and to set up a common review process.

3.3.5 Organizational position

The place of an intelligence unit should be clearly defined in an organizational chart. Accordingly, it is important to identify interactions with other elements of the organization in question. In order that the supplied intelligence products are wholly incorporated into the decision-making process, it is critical that an intelligence unit reports directly to the leader of the organization, or to his or her deputy, and that the unit has direct access to internal sources of information. The organizational structure – whether functional, divisional or a matrix – can facilitate or hinder the work of the intelligence unit. In a functional organization, the intelligence unit should report directly to the chief and be integrated as a special unit in the general services. Members of the unit should have direct contact with the different functional areas, and rules should regulate the flow of information (what must be provided, by whom, to whom, when and how).

In divisional or matrix organizations, a central intelligence unit should be established at the headquarters, reporting to the chief. In line with the size of the organization, intelligence cells, or intelligence correspondents, should be placed in each division and report to the central unit. The chief of this unit and the intelligence officer establishes the rules regulating the flow of information. In matrix organizations, a difficulty can arise due to the double reporting system (functional and administrative).

3.3.6 Unit structure

Data collected from field and desk-based research are conducted. A data bank, easily accessible by those with appropriate authorization, supports the work of analysts. This data bank plays the role of organizational memory and should gradually evolve towards a comprehensive knowledge base.

An *analysis section* deals with data and develops the intelligence products to be delivered to decision-makers, or serves as general information for those who "need to know."

The *deployment of field sources* and related monitoring should be considered a function of the intelligence unit, or be handled by another section of the organization. Generally, field sources are collaborators performing their normal duties, which are required to supply specific or ongoing information according to the collection plan established by the intelligence officer (IO)/operation chief (OC). Nevertheless, it may be the case that specific information has to be collected by a covert agent or other special agent. The problems involved in such operations are difficult, and it is thus necessary to closely monitor and follow-up this type of action; the intelligence unit can be tasked to undertake this follow-up activity. In a large organization, such tasks may be handled by a specific department.

Case Study: Classical structure of an intelligence service

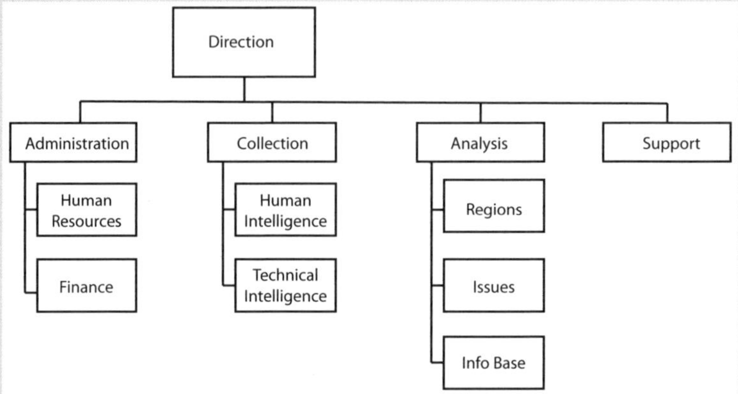

Figure 3.9 Classical structure of an intelligence unit.

The German approach differentiates collection and analysis; in US, structures tend to bring these functions together by geographical area.

3.3.7 Tasks and profiles of intelligence unit personnel

The intelligence cycle encompasses the different tasks that have to be mastered if one aims to establish an efficient organization. These tasks can be detailed according to their different contributions to the intelligence process. Annex 2 at the conclusion of this Chapter provides examples.

3.3.8 Mechanisms of coordination

Within the unit

Internal unit processes should not differ from mechanisms used in other units, such as tasking and coordination meetings.

With the decision-making process

The most important interactions concern the integration of the intelligence function into the decision-making process. The Introduction and Chapter 2 described the role of intelligence in the different steps of the decision-making process.

With external partners and stakeholders

The framework of external coordination must be established taking into account possible legal requirements.

Coordination can result from the

- exchange of liaison personnel;
- establishment of common meetings or working groups; or
- establishment of common centers, either on a regional or a national basis.

3.4 OPERATIONAL STRUCTURES

Turning to structures mobilized in the context of specific cases, uncommon events require a form of crisis management for which normal police structures are not adequate.

3.4.1 A fundamental principle: graduation

A key principle to be followed in the design of specialized operational structures is the "graduation principle." One can draw an analogy in this context to Russian dolls, with the smallest doll representing the core structure, and the larger dolls representing the different elements that can be added according to the requirements of the case. For example, the control and command center (CC) of a police force provides the permanent core element, which can be reinforced according to the size of the event to be managed; if necessary, specific structures should take over the overall management of the case. The following example illustrates one possible way of handling the different situations (normal versus particular, particular versus crisis). Differentiations between such cases can be made on the basis of relative complexity and the resources necessary for an effective response. Command structures and the distribution of responsibilities should be adapted accordingly. Three levels of intervention exist: normal, particular, and crisis; and four

Table 3.10 The graduation principle.

Case level	Patrol	Command center	Duty officer
Normal	Can handle the case; Informs CC.	Informed by patrol; May request more information or search for supplementary information.	
Particular	Cannot handle the case alone; Takes first measures Informs CC; Requests reinforcement; Indicates point of contact.	Assesses situation; Takes first measures at his/her level; Sends reinforcement; Manages information retrieval and dissemination.	Informed; Ensures that adequate first measures have been taken and information disseminated; Takes complementary measures if necessary.
Crisis	Takes first measures; Informs CC; Requests reinforcement; Indicates point of contact.	Assesses situation; Takes first measures; Informs and deploys reinforcements; Conducts operations as long as responsibilities have not been transferred; Manages information retrieval and dissemination.	Informed; Ensures that adequate first measures have been taken and information disseminated; Decides on mobilizing specific structures; Assumes interim command until specific structures take over.

types of stakeholders: police officers (patrol, detective), command centre (CC), duty officers, and crisis structures. The first three are fixed; the last must be mobilized. Table 3.10 synthesizes the process.

3.4.2 Specific structures

There are two categories of "specific structures" that can be mobilized. The first is composed of specialists or specialized units that are placed at the disposal of the responsible leader (such as a bomb squad, SWAT team or negotiation team); the second consists of full command structures that will take over the role of crisis management (with the specialized teams then integrated into the overall 'crisis' resources). In such situations, the flexibility ensured by the graduation principle offers great advantages to the chief of operation.

Case Study: The advantages of gradual mobilization

Upon taking over the command of a police force (by one of the authors of this book) at a time when disaster-crisis structures were tested in what was fortunately just an exercise, though on a very large scale (more than 100 participants from different agencies – police, fire department, health department – were involved, as well as technical support). The exercise scenario was a chemical explosion, with a significant amount of wounded and killed individuals; the weather contributed to the difficulties faced, as the heavy downpour of snow made the traffic hazardous. An assessment of the situation led to the activation of the Disaster Plan. Once the decision had been made, however, full deployment of the plan raised a whole series of logistical and operational problems. The debriefing showed that greater flexibility would have reduced many of the problems encountered, as it would not have been necessary to deploy certain resources (at least not immediately). On the basis of this negative experience, the chief of operation of the Disaster Plan reviewed the entire mobilization process and introduced much greater flexibility, allowing mobilization of resources tailored to the situation in question. Further events and exercises confirmed that the new structure was more appropriate than the one it replaced.

3.4.3 Progressive deployment of crisis structures

No single best solution exists in the context of crisis planning; depending on the scenario in question, one may or may not be able to gradually mobilize crisis structures. What remains important is that one has the freedom to decide. Table 3.11 describes a progressive process that is just one of the possible ways for organizing such structures.

Table 3.11 Progressive crisis planning.

Situation	Mobilization	Remarks
Serious or uncommon events that cannot be managed by normal ways and means.	*First step* After assessment of the situation by the CC and duty officer, mobilization of the first layer of crisis management staff (CMS).	*Location*: Crisis Control Room. *Participants*: Operation chief/or deputy; Chief or deputy of uniformed police, criminal police, support services; Intelligence cell; Telecommunications; Specialists as needed (press officer, logistics officer). *Mission*: Assessment of the situation and deciding on the resources to be deployed; *Locations*: CMS: Crisis Control Room First Responder CMS: place of event
	Second step (decision has been taken to mobilize the First Responder CMS and to allocate adequate resources); The personnel of the CMS is completed if necessary; The Front CMS is deployed at the location of the event and takes over the management of the crisis there.	*Participants*: As needed by the management of the mission. *Missions* (parts of) CMS: Overall management and coordination with political authorities, other agencies/relations with the media. Front CMS: Management of the event at the place of intervention.
Remarks	Participants and allocated resources will depend on the size and type of event. In the case of a large disaster, several agencies will be integrated. There may be several FCMS if the event covers a large area; for instance, in case of a natural disaster affecting several villages and cities. The individual in charge of the FCMS is responsible for the service directly concerned by the event, such as the chief of the fire department. The decision process is iterative between CMS and FCMS. Intelligence cells are established at each level of the command structure. Significant emphasis should be placed on the mobilization procedures, including: the personnel to be mobilized, the priorities of mobilization, and the procedures and means of mobilization.	

3.4.4 Some operational structures

This Section considers three examples of structures designed for handling specific or uncommon events.

3.4.5 Planned events

A police force may be required to organize security for a large event, such as an international conference held in a town situated within its jurisdiction, or at a sporting event that could be troubled by violent incidents. Two police organizational structures may be appropriate in such situations, the first for the preparation phase and the second for the execution phase. While the focus here is on police forces, it is clear that either links with other agencies and authorities must also be established, in setting up channels of communication, or integrating their representatives into the structures described.

Planning structures

Planning staff cells are structured around the problems to be tackled: information and intelligence, tactical deployment, logistical and technical problems, and media.

Information is divided across two different branches (Figure 3.12): Intelligence, which collects and analyzes all data pertinent to threats faced, and information, which collects and handles all other elements that may be needed to make various kinds of decisions (*e.g.* information on possible premises for accommodating security personnel).

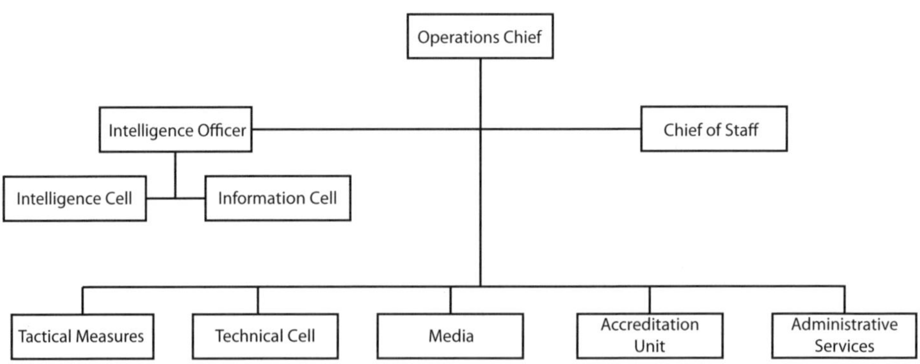

Figure 3.12 Planning structures.

Command structures

Planning structures evolve into command structures when the operational phase begins (Figure 3.13). In general, two different levels are considered: the central command post, which is normally, located at police headquarters, and in a single (or in some cases several) decentralized post, established at or in close proximity to the location where the incident takes place. Figure 3.13 represents an example of a possible structure for planning an operation.

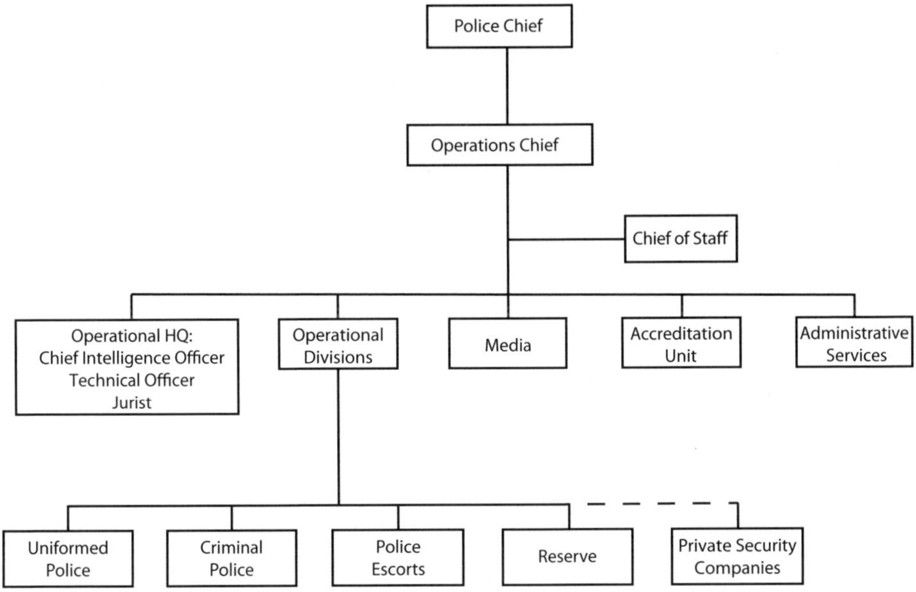

Figure 3.13 Command structure.

Disasters and other larger events

In some cases, the nature of the problem will be apparent from the beginning. A train accident with many casualties will trigger an immediate emergency response. In other cases, the risks involved may not be identified at once, or may not necessitate a full deployment of the plan. Reinforcements could be brought in as required by the situation. In such cases, the police deployment would soon be reinforced by other agencies. The flexible mobilization of resources *and the existence of prepared and trained contingency plans are primary components* determining the efficiency *of the measures taken.*

Hostage management crisis

The structures assembled to respond to a hostage crisis are another example of the adaptation of core police structures. A centered headquarters staff with the chief of operation and his or her direct subordinates, a decentralized on-site command post and the deployment of resources for managing the problem are generally mobilized.

The core of the CMS remains the same, but can be completed by a psychologist, a medical doctor, a legal adviser and/or other specialists. The political nature of the event may have to be addressed and liaison with relevant authorities established. State liaison officers would certainly be detached to the CMS. The mobilization of further staff members and the reinforcement of resources deployed at the location of the incident depend on the situation.

The principal adjustment to the basic structure occurs at the incident site; besides the command post, a negotiation cell is brought in to get in touch with the hostage takers and conduct negotiations. A negotiation cell is usually separate from the front CMS, with no bilateral contact. Decisions are only made at the level of the chief of operation at the CMS.

3.5 OPERATIONAL STRUCTURES AND INTELLIGENCE

The issue of information collection was addressed in Chapter 2. This Section will look at the way data is handled when provided to the intelligence unit of the CMS; that is, dealing with intelligence in the context of a specific structure, working to solve a well identified problem. The focus of such intelligence work is thus on collecting data directly connected to the ongoing operation and analyzing the accumulated material in order to support the chief of operations' decision-making. While ongoing intelligence work integrated into the normal structures of a police force will provide data and analysis for operational decision-making, the unit will also supply analysis of past events and preventive actions to reduce identified risks with a view to future developments. Thus, the type of analysis provided will differ from the very "matter of fact" analysis in operations.

Three problems will be addressed here: how does one organize the flow of information within an intelligence unit; how should one establish the CMS control room; and how can one integrate intelligence products into the decision-making process? The operational role of the intelligence officer and the system of reports will be described further in Chapter 5.

3.5.1 Treatment of information within the intelligence unit

The operational chain normally assumes the following form:

- Data entry (all data or information derived from communications is recorded, indicating the source, time and origin).
- Evaluation of the quality of information.
- Recommendation of measures based on information evaluated.
- Inscription of information in a logbook (journal of events).
- If necessary, mapping of information.
- Dissemination of information according to established procedures.
- Registration of information.

To perform this set of operations, the operational intelligence unit can be organized as in Figure 3.14.

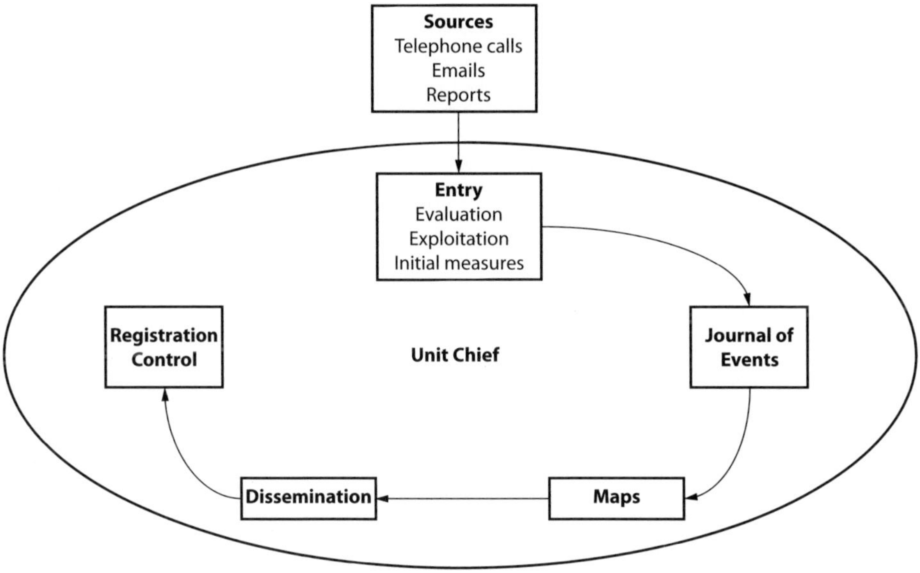

Figure 3.14 Operational intelligence unit.

3.5.2 Control room

The establishment of a control room is necessary in the context of operations of a size and duration that implies the coordination of different entities and shift-work in order to ensure the continuity of command. In such cases, it is essential to maintain an overview of the operation and to ensure that information is adequately transmitted at each change of shift of headquarters staff.

One should find three types of information in the room (displayed on the walls, for instance), summarized in Table 3.15. Figure 3.16 provides an example of the Compstat conference room.

Table 3.15 The three types of information for the control room.

Operation in process	Mission; Operational intention (how to execute the mission); Charts (chain of command, deployed resources); Maps; Telecommunications, liaison contacts; Specific measures (what, who, when, status); Additional information.
Information	Weather; Deadlines; Next briefing; Additional information.
Organization	Shift plan; Organization of the staff; Other information.

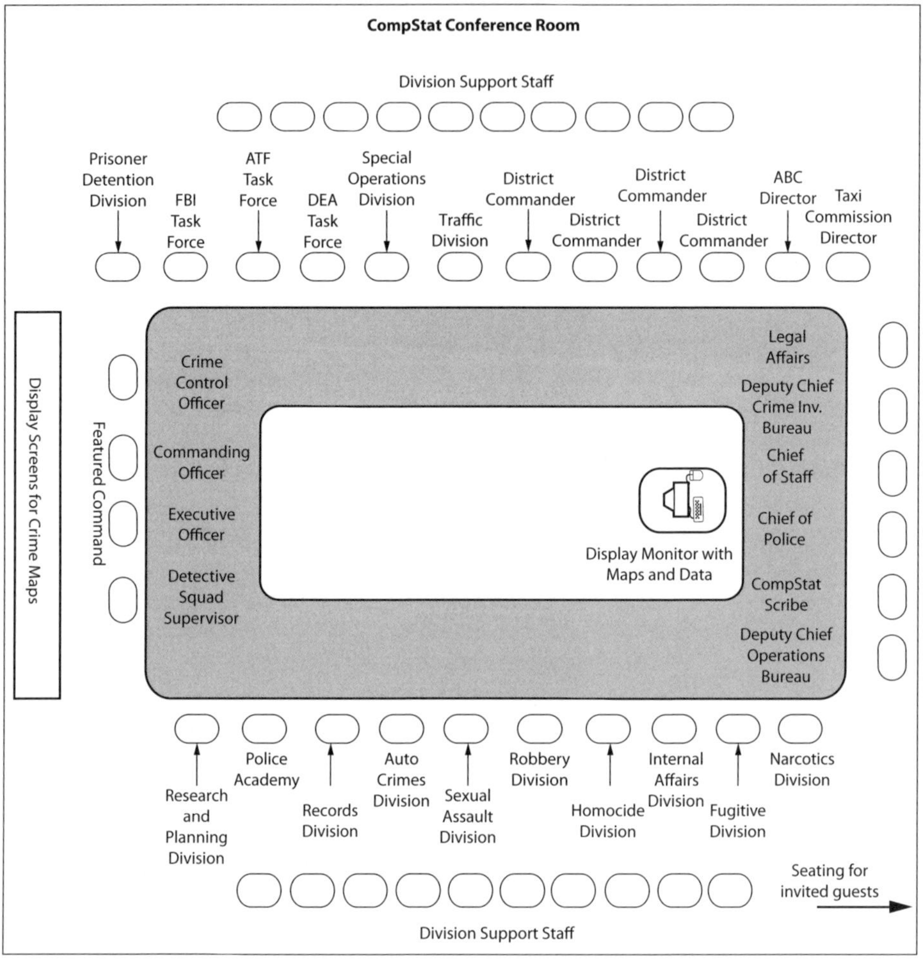

Figure 3.16 Compstat conference room (reproduced with kind permission of the author[5]).

[5] Shane, J., "Compstat Implementation", *FBI Law Enforcement Bulletin*, Quantico: FBI Academy, June 2004.

3.6 SUMMARY AND KEY POINTS

- The implementation of a strategy or decision is achieved via structures, which can be either administrative or operational.
- While structure should logically follow strategy, frequently the formulation of a strategy must respect existing structural constraints.
- There are many types of organizational structures; the principal structures include the functional organization, the divisional organization and the matrix. Each involves advantages and disadvantages in terms of the relationship between centralization and decentralization, delegation of authority and control, and flow of information.
- One of the main factors of success regarding operational structures lies in the degree of flexibility surrounding mobilization in terms of time and resources; in the Russian dolls example, sufficient flexibility is achieved if a core element can be completed according to the type of situation faced.
- Intelligence personnel and resources must be structurally integrated into the organization and operations. Proximity to the decision-maker and the flow of information are two key elements to be taken into account in the design of intelligence structures.
- Much attention must be devoted to the description of relevant processes and of the job descriptions of the members of the intelligence unit.

ANNEX 3.1
ADMINISTRATIVE ORDER

The example below (Figure 3.17) provides a general framework for establishing an intelligence unit by identifying all the aspects to be considered. Additional details will depend on the specific organizational context in which the intelligence unit is set up.

Administrative order
1. Mission • General mission 2. Tasks and general duties of the intelligence organization • Tasks • Structuring the intelligence and risk analysis systems – Strategic – Operational – Tactical • Basic methods • Common sources 3. Organization of the intelligence unit 3.1. Geographical levels • Central / strategic • Regional / operational • Local / tactical 3.2. Organization and duties at the strategic level • Mission and tasks • Structure and authority • Link to decision-making 3.3. Organization and duties at the operational and tactical levels • Mission and tasks • Structure and authority (operational and tactical level) • Link to decision-making 3.4. Relationships between the different levels 4. Specific criminal intelligence operations and targets • Different types of information • Who has the authority to order operations, define targets (all levels), require information/intelligence? • Monitoring and reporting 5. Collecting and handling of information • Collecting of information a. Duties of all police officers b. Duties of intelligence personnel • Evaluation of information • Use of databases

Administrative order (cont.)
6. Analysis and reporting • Cooperation reports • Intelligence analysis products • Dissemination of products • Description of decision-making
7. Assessment of results
8. Platforms for the exchange of information • Rights to give information to other national authorities (all levels) • Cooperation with investigators • Use of liaison officers
9. Training and education • General needs • Requirements of personnel: profiles and job descriptions • Selection process • Responsibilities • Program • Cooperation with other authorities
10. Internal communication

Figure 3.17 Example of the establishment of an intelligence unit.

ANNEX 3.2
POSITIONS IN AN INTELLIGENCE UNIT

Table 3.18 Summary of positions in an intelligence unit (cont.).

Position	Contribution	Specific tasks	Profile
Intelligence officer/ operation chief	General responsibility and accountability.	Stating clear requests for the collection of intelligence.	
Director of the Intelligence Unit Intelligence officer (or manager) (the example given here relates to an organization in which the manager fulfills two distinct functions: managing the unit and serving as intelligence officer)	Ensuring the effectiveness and efficiency of the intelligence unit; Adding meaning and weight to the analytical techniques and products[6]; Contributing to assessments.	Managing the intelligence unit (setting objectives, implementing policies and procedures, allocating resources, coordinating, motivating, selecting qualified personnel, training, etc.); Developing and presenting the intelligence products; Disseminating information in appropriate forms, and in sufficient time; Developing proposals in his or her area of responsibility; In each state, developing proposals relating to the collection and dissemination of new information; Operational tasks.	Manager with in-depth knowledge of intelligence work and of the decision-making process.
Collection (field officers to supply information and desk researchers to obtain information from different sources that must be clearly stated)	Providing the information necessary to assist decision-making.	Determining information to collect; Defining and training sources; Defining methods and means of collection; Defining and overseeing security requirements (physical, procedural, human).	Field officers with an ability to proactively recruit and deploy human sources, and handle technical collection.

6 The National Intelligence Model booklet p. 21, published by National Criminal Intelligence Service, London www.policereform.gov.uk/implementation/natintellmodel.html (last accessed on June 23, 2010).

Table 3.18 Summary of positions in an intelligence unit (cont.).

Position	Contribution	Specific tasks	Profile
Collation Data management section	Selecting significant information and organizing registration; Organizing documentation; Establishing a knowledge data base.	Inputting data; Providing access to those authorized; Offering research support to analysts.	Skilled in data management.
Interpretation Analysis section	Transforming information into intelligence.	Analyzing; Explaining, clarifying, highlighting; Developing products.	Analytical capabilities.
Dissemination (Normally not a special unit unless part of a very large organization. Preparing bulletins, etc.)	Transmitting information in an appropriate form and in due time to those who need to know.	Identifying those who need to know; Defining form, frequency, and means of dissemination; Delivering presentations.	Capabilities in the field of communication.
Feedback	Monitoring the process and making necessary adjustments.	Control.	Administrative skills in the field of intelligence.

Chapter 4

Power and Leadership

Objectives

While structures are important, men and women remain fundamental to the success of an organization. Though their actions may be facilitated or complicated by structures and other aspects, the "human factor" will continue to be a vital aspect of organizational success or failure. Members of an organization will learn how to use relevant structures in order to foster their own goals and interests; at the same time, inefficiencies and conflicts may result from clashes between personal interests and goals and collective aims. This chapter examines the central issues of power, leadership and relational conflicts from an organizational perspective. The rational decision-making process described in Chapter 2 may be affected at each of its phases by these phenomena: power and leadership.

4.1 WHAT IS POWER?

Chapter 1 developed the notion of a system, defined as an "ensemble of elements and actors, interacting mutually and with their environment, and organized to achieve certain objectives." In the security system, for instance, the chief actors are the military, the police, the border guards, and private security companies.

In this context, an organization can be described as a social entity (interactions between its members), rationally organized (management, structures, procedures), accomplishing activities (mission) in a framework whose boundaries are fixed (territories, service line, customers), and pursuing clearly identified goals (strategy) in an environment with which it interacts. An organization can also be seen as an arena where competition takes place between members whose interests and objectives may differ from the collective. In order to achieve their personal goals, individuals may exploit the specific advantages linked to their organizational position – if they possess a key specialization – provided these are relevant to the opportunities they wish to exploit. Hence, an organization should not be seen as particularly rationally driven; inefficiencies and conflicts must be reckoned with. Power must therefore be exercised in order to avoid special interests prevailing over organization goals. Power can be employed positively for achieving organizational goals. An analysis of an organization must take into account the different actors involved and try to understand their interests and objectives, which strongly influence organizational functioning. Taking the example of the criminal division of a police force, one can draw the following illustration. At the bottom, the individual detective will aspire to promotion; at the unit level, the chief will also seize every opportunity for promotion, but may also fight to increase the resources allocated to his or her unit and oppose other unit chiefs who pursue the same objective. All will rally, however, to enlarge the autonomy and advantages of the criminal branch within the force and all police – detectives and uniformed officers – will unify when advantageous, in order to obtain more benefits from political authorities.

Power, even in a hierarchical organization like a police force, is never unlimited and cannot always be identified by looking at an organizational chart. Power can be defined as the ability of certain individuals or groups to act over other individuals or groups. Conversely, the ability of certain individuals or groups to resist the power of others in an organization is also a form of power acquisition, as it will confer upon them a kind of bargaining power. Negotiation is therefore a key concept, and power can be measured according to the more or less strong position it confers to those who sit around a negotiating table.

According to French and Raven's classification, there are five types of power: reward power, coercive power, legitimate power, expert power, and charismatic power.[1] One should add the power of nuisance, because if an individual or group can harm the functioning of an organization, they will obtain a power of negotiation vis-à-vis the management. The consequence of such a situation is that an organization chart does not accurately display the relationships within an organization; members who may not have a senior position may, however, have a strong influence because of their charisma, their relations, or their expertise; legitimate power can hence be weakened or even kept in check.

4.1.1 Power within an organization

Mintzberg studied the distribution of power within an enterprise.[2] He differentiates between six categories of personnel; each one owns a certain power that influences the operation of the organization: general direction, management, skilled operators, unskilled operators, analysts and support personnel. Applying his grid to a police organization, one can establish the following simplified structure of power (Table 4.1).

---

1 John. P. R. French, Jr., and Bertram Raven, "The Bases of Social Power" in: Dorwin Cartwright and Alvin F. Zander (eds.), *Group Dynamic*, Harper and Row, New York, 1960, pp. 607-623.

2 Henry Mintzberg, *Le pouvoir dans les organisations*, Editions d'Organisation, Paris, 2003, pp. 64-69.

Table 4.1 Distribution of power in a police organization.

Function	Types of power	Influence
Chief	Legitimate (high); Reward and punishment (partly direct, partly indirect, depending on decisions at political or judicial levels); Expertise (variable); Charisma (variable).	Legitimate power confers an influential position. His or her expertise, charisma and control of information will add or diminish the level of influence enjoyed. His or her relations with political authorities and the outside world will also play a role.
Officers	Legitimate (high); Reward and punishment (generally indirect but important); Expertise (technically high); Charisma (variable).	Depends mainly on the recognition that their subordinates give them based on their expertise and charisma.
Police officers	Mass power rather than individual; Power (legitimate) more directed towards the outside world than within the organization;	Their power may come from their more or less large autonomy (detectives have more than uniformed police officers) and of their aggregate power, which they can apply in negotiations.
Civilian	Expertise power.	Their influence will depend on their area of expertise; that is why their power will instead be attached to an individual accomplishing a specialized function rather than to a group. A computer specialist may play a key role but a waiter at the police restaurant will not.

4.1.2 External power and its influence on an organization

It is now possible to consider the external influences that can be exerted on an organization by the following actors (Table 4.2).

The practical consequences of this analysis are that each manager should endeavour to understand how he or she can influence others and be influenced by them. For that purpose, a map of influences can be drawn, and actions prepared, to capitalize on favorable dynamics and reduce the effect of negative relationships. Active measures can consist of concluding alliances and establishing contacts with key individuals or groups. The search for information and environmental scanning in order to detect potential risks should help a manager to take preventive or counter-measures for assuring his or her power.

Table 4.2 Actors involved in external influence of power.

Actor	Types of power	Influence and potential problems
Political authority	Legitimate.	Direct through its capacity to act at the strategic, legal, and administrative levels. Strong through budgetary procedures and the allocation of resources. Potential conflicts due to the risk of interference in the operational sphere.
Civil society	Indirect through the values it conveys.	Important for a police force. Potential conflicts when values of the organization differ from those of the society.
Public opinion	Indirect through surveys, complaints, letters to newspapers, demonstrations.	The influence on a police force/an organization will be mainly indirect, as the concerns expressed via public opinion may be taken up by political authorities and the media.
Pressure groups, lobbies	Indirect through advocacy at the political level, and via media and public opinion.	Influence depends on the cause and the clout of the lobby. Potential conflicts due to the political interference resulting from such pressures.
Media	Indirect power through its influence on politics, opinion leaders, public opinion.	Risk of the police/organizational agenda being dictated by the media according to the attention devoted to certain causes. A feeling of insecurity is for instance largely provoked by reports on crime presented on TV and in newspapers.
Trade unions	(as above) Links with their correspondents within an organization.	(as above) Potential problem within the organization if a case is taken up by internal unions.
Organizational partners	Indirect if decisions made by them have consequences for the organization (for instance, police may have to resume duties relinquished by another member of the security system. In the economy, a company may depend heavily on one or a few customers or suppliers).	The influence and consequences depend on the type of relationship and on the degree of autonomy enjoyed.

4.1.3 How to acquire power?

What are the principal means of obtaining influence? Specific skills – as far they are important for the organization (think of computer specialists) – access and control over information, and access and proximity to the decision-maker (for example, a chief's secretary, who can act as a gatekeeper) are elements conferring power to their holders, irrespective of their position in the formal organizational hierarchy.

Building up a reputation (of expertise, of success, of being well connected) is another powerful means; but perhaps the most effective way of acquiring influence over others is to create a relationship of dependence. This can hinge on the capacity of the superior to provide the subordinate with the necessary resources to succeed. Hence, the chief's subordinates will evaluate him or her according to his or her capacity to obtain new resources. Misunderstandings in police forces between chiefs and police officers frequently arise from the fact that the objectives and interests of both parties are different. Police officers expect more resources – they create new possibilities of promotion – and a chief, restricted by financial constraints determined by political authorities, will often have to search for alternatives ways of bolstering the means available. He or she may have to resort to measures of restructuring and personnel will only reluctantly accept downsizing. To manage such situations, a chief will have to maneuver with dexterity. He or she must on the one hand comply with the allocated budget and on the other hand focus on maintaining a positive climate within the force. How can one achieve such a result? Transparency, communication and contacts are primary components of success in this type of situation.

Nevertheless, subordinates can also create a relationship of dependence with their superiors. Their power may stem from their unique expertise or from relations established with influential people outside the force or because they have been chosen to represent others. Union leaders within a police force are important figures in this context. One can thus observe that power is not unidirectional. Legitimate power does not always suffice and real authority must be backed by other factors.

4.2 PROBLEMS BETWEEN HOLDERS OF POWER

Potential problems can arise in the interactions between parts and/or members of an organization. While the focus here is predominantly on a police environment, the same difficulties can be found in other types of organizations. The problem is always the same: conflicts relate to the distribution of power between members of a system or of an organization. Personal or specific objectives and interests may collide with those of the organization.

Three areas of difficulties can be identified: relationships between politicians and operational leaders (in a private company between the Board of Directors and the CEO), relationships between the headquarters and divisions (subsidiaries in a private company), and relationships between divisions, units or members of the organization. An illustration can be made in the relationship between uniformed police officers and detectives or between police forces coexisting on the same territory (in a private company, this type of conflict may be found in the competition between production and marketing).

Focusing first on the tensions that may arise between politicians and operational decision-makers, one can learn from an observation of the wartime relations between politicians and generals.

Case Study: Politicians and generals in wartime

Differences of opinion may relate to the goals pursued by politicians and generals during wartime. For instance, in Gulf War I, General Schwarzkopf wanted to enter Baghdad but President Bush vetoed this move, as he knew that Saddam Hussein's overthrow would destroy the regional balance of power. Another cause of conflict lies in the possible meddling of politicians in what generals believe to be their sphere of responsibilities. Hitler considered his generals as better-paid non-commissioned officers and imposed his own views upon them.

Churchill on the other hand frequently irritated his generals by incessant close questioning and relentlessly querying their assumptions and arguments.[3] There was, however, a great difference between both leaders: Hitler imposed his will without taking into account his generals' opinions, Churchill, even if his style frequently upset his collaborators, instilled within them, as Norman Brook, a member of Churchill's Cabinet Secretariat, noted, "a new sense of purpose and of urgency as it came to be realized that a firm hand, guided by a strong will, was on the wheel."[4] Differences in the use of means will also provoke disagreements between both partners as shown by President Truman refusing General McArthur's request to resort to the atomic bomb during the war in Korea.

Military chiefs may also try to influence politicians. Manipulating information, exerting their expertise, and a policy of *fait accompli* are three of the possibilities they have to restrict politicians' freedom of choice. One famous example was the Schlieffen Plan.[5] This plan did not take into account any political or diplomatic factors and did not contain any alternative solution. The plan was followed and the violation of Belgian neutrality involved led to the entry of Great Britain into the war. The launching of the Schlieffen plan had another incalculable consequence, as it meant that politicians were subordinated to military chiefs when the army mobilized.

---

3 Eliot A.Cohen, *Supreme Command*, New York: Anchor Books, 2003, 95-132.

4 *Idem*, 124.

5 In 1904, France and Britain signed an alliance the aim of which was to encourage cooperation against the perceived threat of Germany. Negotiations also began to add Russia to this alliance. As a result of these moves, the German military began to fear the possibility of a combined attack from France, Britain and Russia. Alfred von Schlieffen, German Army Chief of Staff, was given instructions to devise a strategy that would be able to counter a joint attack. Schlieffen argued that if war took place, it was vital that France was speedily defeated. If this happened, Britain and Russia would be unwilling to carry on fighting. Schlieffen calculated that it would take Russia six weeks to organize its large army for an attack on Germany. Therefore, it was vitally important to force France to surrender before Russia was ready to use all its forces. Schlieffen's plan involved using 90% of Germany's armed forces to attack France. Fearing the French forts on the border with Germany, Schlieffen foresaw a scythe-like attack through Holland, Belgium and Luxembourg.

A third kind of conflict between two closely interrelated parties concerns organizational headquarters (the center) versus divisions or subsidiaries (either in a police force or in a private company). The normal division of tasks attributes to the center the development of a vision and implementation of strategy, management of common resources, and advisory and control roles; the divisions/subsidiaries should contribute and act within the strategy. The search for more autonomy and a better control over their resources or a larger share, even at the expense of other divisions, will often cause difficulties. Decentralization often leads to the search for greater autonomy by the various levels of units; the center may resent its loss of power and try to reaffirm its control through tighter regulation, procedures, and objectives, which may not be aligned with the issues faced by divisions, contributing to the emergence of conflicts.

Tensions can also arise between the different members or groups of an organization. It has been previously noted that individuals or groups pursue objectives that can differ from those of the organization as a whole. To this end, individuals and groups can try to increase their independence and build on their specific expertise and position to gain ascendance over other members or groups in the organization. In each police force, one will find, for instance, some form of conflict between uniformed police officers and detectives. Their sub-culture will be different as well as their professional objectives. One should emphasize that this conflict will appear both in police forces that have separate recruitment processes for each kind of police officer, as well as in those contexts where detectives were former patrol officers. How can such conflicts be addressed in order to avoid too great an impact on the life of an organization?

Concentrating on the interactions between protagonists, one can focus on two types of situations: relations between the center (HQ) and its divisions/subsidiaries, and between the politician in charge of security and the chief of police.

Concerning the first situation, three models of relations can be outlined:

- The HQ is strong and the periphery is weak; this model can be qualified as authoritative. In such a case, the former is able to impose its will and the latter will adhere. There may be no apparent conflict, but questions of how willingly the periphery will engage with the objectives set at the center, and secondly if such a situation doesn't prevent many ideas developed in the periphery to flow to the center.
- The anarchical model in which the HQ is weak and the periphery is distinctly autonomous. In such a situation, the objectives set at the center will be followed by its subsidiaries provided they do not infringe on their own objectives.
- The third model is democratic and based on a cooperative partnership between both parties, which jointly define the strategy and objectives and clearly agree on the division of tasks.

A similar situation occurs in countries with different police forces. Almost inevitably, competition between forces will develop with regards to the distribution of power and responsibility. Examples of such situations may be found in France between the gendarmerie and the National Police, in the USA between the federal agencies and other police forces, in Germany between the *Bundeskriminalamt* and the police of the Länder, in Switzerland between the cantons and city police forces, and so on. Opposition between stakeholders mainly centers on issues of centralization versus decentralization and thus concerns the roles and powers of political authorities and chiefs at different layers of the system. The results depend on the respective power of the protagonists, chiefly the political authorities. The following situations can occur:

- *Status quo* when no protagonist is able to change the system because the balance of power between the different political levels, chiefly national, federal versus regional or local does not allow one party to prevail over the other.
- Consolidation is reached when the balance of power is favorable, for legislative, economic or other reasons, to the national, federal representation. Austria has succeeded in unifying its four police forces because its strong centralization.
- Dilution of power happens when the balance of power lies in favor of the regional or local levels. The municipal police in Belgium have enjoyed the upper hand over the national gendarmerie.
- Coordination is a solution when stakeholders establish compromises, which generally consist of granting a coordinating role to the federal or national level against maintaining the independence of the other parties to which more responsibilities are delegated.

The same pattern may be used to outline the framework of interactions between the politician in charge of security and the chief of police. The following cases can be identified:

- The politician's influence is strong and he or she considers the chief as the executor of his or her policy; if the chief is weak, the relationship will likely be feasible, if the chief is strong, he or she may not accept playing a secondary role and tensions may arise.
- If the reverse is true, that is, if the chief is strong and the politician considers it in their best interest to follow him or her, the relations may remain harmonious, at least, when there are no problems. If difficulties arise, the chief could not expect strong support from his or her superior. On the other hand, the risk of conflict in this type of situation is high, as the chief may be tempted to go beyond his or her competence, in the process damaging the politician's ego.
- When both parties are strong and competent, their relationship may lead to fruitful results. As the clash between Giuliani and Bratton shows (see the case study below), this situation can also lead to destructive fights between egos. The fight for visibility and a public profile may be one of the areas where the interests of both stakeholders can frequently collide. Politicians must be elected, and acquiring notoriety in the media and public is one method of achieving such a goal. When a chief or senior civil servant makes the headlines (in the positive sense), political superiors can be offended and envious.

Case Study: NYPD[6]

Mayor Rudy Giuliani, who appointed Bill Bratton as chief of the NYPD in 1994, forced him out of his job after two years in what was generally seen as a battle of two large egos in which Giuliani was unable to accept Bratton's celebrity. The mayor had made law enforcement his signature issue and Bratton his police commissioner. Crime fell sharply, and both became stars. However, an administration that prized unwavering loyalty to the mayor could not stomach Bratton's celebrity. He dined with celebrities and appeared on the cover of Time Magazine in 1996. He tried to organize a parade for the Police Department's 150th anniversary, prominently featuring himself, but Mayor Giuliani killed the plan. Chief Bratton left the job after just two years of very successful activity.

- If both parties are weak, their relations may be good but policy will lack dynamism and necessary changes will not be achieved.

What consequences can one draw from this analytical framework? Four factors may influence power relationships: the broader environment, the personal qualities of the protagonists, their objectives, and their way of cooperating. Much will depend on pressures resulting from the environment. In times of unrest, trouble and change, when policies must be reviewed and reforms undertaken, tensions will arise. The personal qualities of the actors, strengths or weaknesses of character, and ability to cooperate will be important as well as their capacity to uphold the general interest over their own goals. Methods of cooperation will have to correspond with the profiles of the protagonists. Autocratic interactions will be possible if one is strong; cooperative relations will give good results if both parties are strong and willing to engage.

Ideal conditions for good cooperation are met when the distinction between what is political and what is operational has been clearly defined and accepted, and when the politician and chief have developed sound and loyal relations, including coming to an agreement on the goals to be achieved, the resources to be provided, and the smooth circulation of information.

4.3 LEADERSHIP

It was necessary to discuss the issues of structure and power prior to leadership, in order to understand that the power of leaders is not unlimited within the boundaries set by structures, other actors, and available resources (to name but a few constraints). The best leaders will know how to survive within this "jungle" while unfit ones will constantly bump up against constraints and remain inefficient, leaving them vulnerable to replacement.

---

6 Richard Pérez-Pena. "Giuliani courts former partner and antagonist." www.nytimes.com/2007/03/09/us/politics/09rudy.html (last access October 26th, 2010).

Each member of an organization knows there are good and bad leaders. The question is: what differentiates them? The following section will try to sketch an answer in identifying first the tasks and roles of leaders, and then present some features of good leadership. Before concluding, some examples will be illustrated of actions and behavior of leaders that have led to success or failure.

4.3.1 Managers and leaders[7]

It is important to first distinguish between managers and leaders. Some authors, like Tom Peters, have sketched a contrast between dull managers striving to administer their organization and flamboyant leaders whose vision, enthusiasm and charisma transforms their followers. This debate will be avoided here, as the qualities that a chief must demonstrate depend on his or her ability to match appropriate methods to different situations that arise. The chief must sometimes manage, and at other times lead. Thus, this section will use the terms manager and leader interchangeably.

4.3.2 Tasks and roles

Mintzberg[8] makes a crucial distinction between the tasks and roles of managers. On this basis, it is easier to list the tasks that should be completed by each leader and the roles best suited to an individual according to his or her personal qualities and skills. Each manager accomplishes (should accomplish) five basic tasks: to analyze and set objectives; to organize; to motivate and communicate; to control; and to evaluate and give feedback. In addition, a manager should train his or her personnel and engage in self-development.

Mintzberg divides the roles of a leader into three categories:

- An interpersonal role, where he or she acts as figurehead of an organization and creates or maintains contact with potential and important partners.
- A communication role, where he or she gathers or passes on information.
- A leader role, where he or she acts as an entrepreneur in developing and realizing a vision, as a fireman when addressing problems and crises, and as a resource allocator when arbitrating between conflicting demands.

These differentiations highlight a truth that is not often times admitted: nobody can possess all the qualities necessary to perfectly fulfill all these roles. Therefore, in selecting a leader one must take into account the kind of contribution he or she will be expected to make. On the other hand, a successful leader knows his or

---

7 Tom Peters, *Thriving on Chaos*, London: Pan Book Ltd., 1989; Tom Peters and Nancy Austin, *A Passion for Excellence*, Warner Books Edition, New York, 1985.
Abraham Zaleznik "Managers and Leaders: Are They Different?", *Harvard Business Review*, January 2004, p. 74.

8 Henry Mintzberg, *Le Management*, Editions d'Organisation, Paris, 2003, pp. 32-41, translated from Henry Mintzberg, *Mintzberg on Management, Inside our Strange World of Organizations*, The Free Press, New York, 1989.

her strengths and weaknesses and tries to compensate for the latter by appointing people who possess the skills lacking. The art of building a team consists of assembling the different qualities covering the entire range of described roles.

4.3.3 Some key issues for a leader

A leader must realize that his or her authority is not absolute. Given the extent to which internal and external factors limit power, it is important for a leader to identify and assess the influence of such factors. Secondly, a leader must realize that his or her authority depends more on credibility than on legitimacy. Thus, a leader should strive to increase their credibility through building upon successes, enhancing their expertise, and enlarging their network of relationships.

A leader must acknowledge that the higher he or she rises in a hierarchy, the less he or she leads directly, instead he or she works through others. Hence, a leader must be ready and know how to delegate. Assuming new responsibilities requires viewing problems from new angles. Drucker notes that "the most common cause of executive failure is inability or unwillingness to change with the demands of a new position. The executive who keeps on doing what he has done successively before he moved is almost bound to fail."[9] Therefore, to succeed in entering a new job, means being able to reflect on that job, identify the requirements and honestly analyze if personal skills and qualities correspond with those requirements. In this context, the notion of a leader's contribution should be addressed. A leader must also answer a key question: What can I do that can't be done by another person, and if done exceptionally, would make a real difference to this organization?" A leader's contribution is to be understood as the key activity he or she alone can provide and if not done by him or her would either not be accomplished, or poorly achieved.

Case Study: Contribution

This has been the similar experience of one of the authors of this book; an account of his personal experience of taking charge of 1000 police officers without previous experience in policing. Before accepting the job, he assessed how he could be successful in his new position. For that purpose, he identified what he could and would provide: his management experience obtained under the general direction of a noted enterprise, his operational and command skills acquired in the army (he was in command of an infantry battalion at the time), and his background in political science and economics. Each allowed him to define three areas on which to focus: management of the organization, operational aspects, and modernization of the force. He thus concentrated his efforts in developing a vision and translating it into a strategy and plan of action (supported by a new communication policy) to achieve these goals.

---

9 Peter Drucker, *The Effective Executive*, Pan Books, London, 1970, pp. 59 and 61.

A leader's contribution will always have to relate to three major areas: achieving results; embodying, communicating and constantly reaffirming organizational values; in addition, developing the skills of personnel, especially those who he or she will lead. Promotion is a key factor for success within an organization. In the context of police promotions, poorly defined expectations can lead to failures. Very frequently, a police officer is promoted based on the good job he or she has done or on the basis of seniority. Often, training for recently promoted officers is also inadequate.

Case Study: The need to delegate

To become a captain in the Swiss army, one has to go through advancement courses, which focus on tactical and technical matters without addressing the administrative issues of managing a company. When newly promoted officers take charge of a unit, they often encounter some difficulties in this area. The experience of one of the authors, however, allowed him to define his expected contribution and identify what he had to focus on and what he had to delegate. The lessons learned on this occasion proved valid for the remainder of his army and professional career.

Returning to the situation in a police force, acute problems can be provoked by errors in promoting people without taking into account the qualities and skills they will have to master in their new responsibilities. Consider, for example, an excellent drug investigator who is promoted to head of the drug unit. In his or her new position, he or she will have to demonstrate management qualities that were not necessary in the previous job. Failure may then be caused by the new manager simply continuing doing what he or she previously did and discarding the tasks attached to the new function. He or she would thus anger the members of his or her team in meddling with their work and frustrate them because he or she does not fully perform the role devoted to the head of the unit; tensions would soon appear. Three lessons can be drawn here: firstly, the quality of the job done at a certain level does not imply that the person has the necessary quality to move up the management ladder; secondly, selection and promotion systems must take into account the profiles of the positions to be filled, and assess the candidates accordingly; thirdly, future leaders must be prepared, through adequate training programs, for their new responsibilities.

In this context, it is essential to understand that the mix of qualities and skills needed for achieving success in a position changes according to the tasks and the level at which they are accomplished. Skills can be divided into three categories: technical, human and conceptual. The relative importance of these skill sets will change according to an individual's position in the organization. Figure 4.3 illustrates the evolution along the course of the career of an officer.

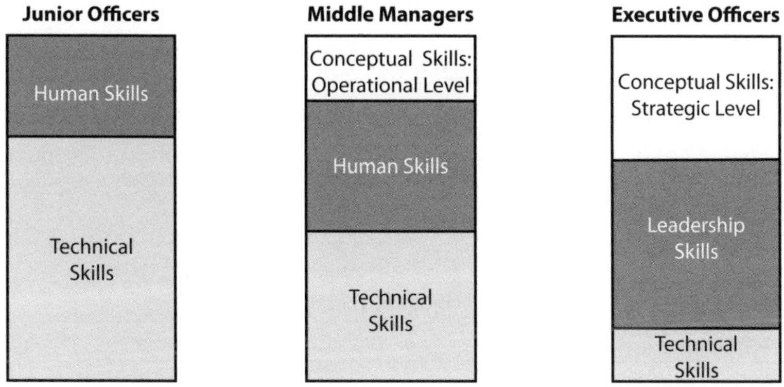

Figure 4.3 Different qualities according to the level of responsibility.

What good leaders do

Drucker[10] notes that a good leader answers the question "What must be done?" and sets priorities accordingly. One could also find the mark of a good leader in the fact that he or she has managed to establish a team that regroups people with different skills. General Schwarzkopf,[11] who commanded coalition forces during the first Gulf War, contends that a good leader must:

- fix what is wrong, not what works well;
- establish clear goals;
- set high standards;
- let others do their job;
- allow room for mistakes;
- when placed in command, the ability to take charge;
- do what is right and what is morally correct.

4.3.4 What causes success or failure?

Experiences in private companies show that the following factors strongly influence the success or the failure of projects (Table 4.4):

[10] Peter Drucker, *The Effective Executive*, Pan Books, London, 1970, chapter 1, 1-25.

[11] General Schwarzkopf stated these principles in the conference he gave in March 1992 in Winterthur, Switzerland.

Table 4.4 Factors that influence success of projects.

Factors of success	Factors of failure
Rigor in the analysis of the problem.	No questioning of traditional rules.
Flexibility in the use of methods.	Use of the same methods in different situations.
Tenacity in respect of principles.	Lack of attention to detail or incapacity to go over details.
Development of a vision.	No sense of direction.
Ability to act.	Inability to handle issues rapidly.
Capacity to communicate a vision.	Poor communication.
Capacity to self-assess.	Inability to perceive weaknesses or take them into account.
Humility.	Arrogance.
Perseverance.	Retaining inefficient staff.

4.3.5 Information, decision and leadership styles

It has been shown that one factor of success consists of grasping the characteristics of new situations and being able to adapt to them. Adaptations may concern the ends (adapting goals), the ways (adapting methods), or the means (adapting resources). Leadership style represents an additional factor that can be linked with the organizational environment and personnel. Beyond the mechanics of the decision-making process, the personality of the leader may greatly influence his or her style of leadership. For instance, what is his or her attitude towards risk? What is his or her degree of self-confidence, or confidence in others? The kinds of problems he or she has to address will also have to be taken into consideration. The following section will consider two important elements concerning leadership style.

The handling of information and options by leaders[12]

Leaders will differ in the way they approach two important elements of the decision-making process: information and options. As a means of considering leadership style in the context of these two aspects, it is useful to construct the following Tables 4.5 and 4.6:

Table 4.5 Decision making process and information.

Necessary	Means that decisions will be made when the level of information is judged sufficient.
Maximal	Means the leader tries to obtain all possible information before deciding.
Single	Means the leader decides alone.
Multiple	Means the leader seeks advice before deciding.

[12] Kenneth R. Brousseau, Michael J. Driver, Gary Hourihan, and Rikard Larsson, "The Seasoned Executive's Decision-Making", *Harvard Business Review*, February 2006.

Table 4.6 Decision making process and leadership style.

Decisive	The mix of these two dimensions form the basis of four different styles: Focus is placed on the speed, efficiency and coherence of a decision. The implementation adheres to the decision made.
Flexible	Focus is placed on the speed of the decision, but the implementation is flexible.
Hierarchical	A great volume of data is analyzed and experts are consulted.
Integrative	A framework for the solution is outlined and different options evaluated.

The decision-making style will differ according to the leader's position, in particular as he or she moves away from the field, depending instead on information provided by subordinates. As a result, senior leaders chiefly use the flexible and integrative styles outlined in Table 4.5 above. Another view of this approach is shown in Figure 4.7.

	Necessary information	Maximum information
Focused on one option	**Decisive**	**Hierarchical**
Considers many options	**Flexible**	**Integrative**

Figure 4.7 Leadership styles.

Leadership styles

The Managerial Grid of Blake and Mouton[13]

The Blake and Mouton grid has contributed significantly to the differentiation of decision-making styles according to the focus established by the leader. The authors have developed a matrix that takes into account the 'human' dimension (concern for people) and the 'result' dimension (achievement of objectives). One can deduce from this grid five leadership styles (Figure 4.8).

[13] Robert Blake and Jan S. Mouton, *The New Managerial Grid: Strategic new insights into a proven system for increasing organization productivity and individual effectiveness, plus a revealing examination of how your managerial style can affect your mental and physical health*, Gulf, Houston, 1978.

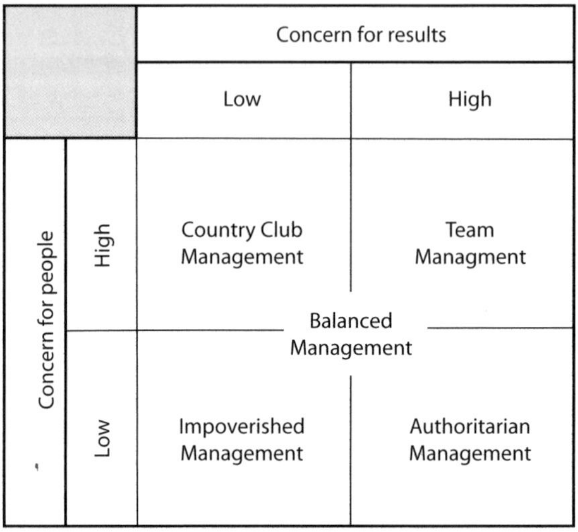

Figure 4.8 The Blake and Mouton grid.

Problems occur when the style applied does not match the situation and/or the profiles of personnel. Crisis management is different from planning management. Scientists cannot be led like soldiers; university professors like line workers; detectives like uniformed police officers; veterans like recruits.

Situational Leadership Theory, Hersey and Blanchard[14]

It was observed earlier that an inability to alter methods in light of changing circumstances could contribute to failure. The same is true of leadership style, not only during the decision-making process, but also in the treatment of personnel. While the Blake and Mouton grid is engaging, it is overly static in considering only two dimensions relating to the leader. Its weakness lies in the fact that the characteristics of the individuals or groups being led are not taken into account. The profiles of these people cannot be dismissed; neither can the nature of the situations in which leadership is applied, nor the leader's character. That is why the further developments suggested by Hersey and Blanchard present an interesting perspective. The situational leadership theory developed by Hersey and Blanchard supplies an instrument that allows one to match management style to a situation and to the personnel involved. The theory takes into account both four styles of leadership, as well as four levels of maturity. A collaborator's maturity is defined according to his or her competency and motivation.

The authors identified different elements linked to the individual that had to be considered in choosing a style of leadership. These criterions were degree of control; degree of support and motivation; degree of development of the individual based on level of competency and motivation. On this basis, they defined four profiles for personnel.

[14] Paul Hersey and Kenneeth H. Blanchard, *Management of Organizational Behavior*, Prentice Hall, Englewood Cliffs, N.J., 1977.

- MI Unable and unwilling or insecure.
- M2 Unable but willing or confident.
- M3 Able but unwilling or insecure.
- M4 Able and willing or confident.

With a focus on the leader, they drew four profiles based on the two dimensions identified by Blake and Mouton: relations and results (Figure 4.9).

		Task Behaviour	
		Low Guidance	High Guidance
Relationship Behaviour	Supportive	**S3** **Participating Style**: The leader shares ideas and facilitates decision-making. The focus is primarily on the relationship.	**S2** **Selling Style**: The leader explains decisions and provides opportunities for clarification. The focus is on the result and on the relationship.
	Unsupportive	**S4** **Delegating Style**: The leader delegates responsibility for decision making and implementation. The focus is neither on the relationship, nor the result.	**S1** **Telling Style**: The leader provides specific instructions and closely supervises the work. The focus is on the result and not on the relationship.

Figure 4.9 Situational leadership.

Hersey and Blanchard concluded by noting that a good leader is one who adapts his or her style of management to their collaborator's maturity.

Table 4.10 Grid showing leader-personnel interaction.

S1/M1	The leader states clearly, what he or she asks the follower to do: what, how, when and where. The communication process is one way.
S2/M2	The leader explains and offers opportunity for clarification; a two-way communication process is established.
S3/M3	The leader shares decision-making through two-way communication.
S4/M4	The leader delegates the task to be done and supervises it.

Table 4.10 may help a leader to define how he or she is going to deal with personnel. The grid also assists in understanding that the level of maturity of an individual is not constant. When a new position is filled or when new tasks are given, the individual has to adapt to new challenges. Hence, he or she could recede from level four to level three. His or her superior must also then change their leadership style accordingly.

At the level of the organization, the grid supplies important information as how to adapt the style of leadership in order to adequately support the policy. For instance, if a police force wants to successfully implement a community-policing concept, it has to empower police officers. They are the ones in direct contact with the citizens and they must be able to directly address various problems and issues. Therefore, the styles S3 and S4 should be used. This has important consequences for the organization, the training and the system of control of the police.

4.4 SUMMARY AND KEY POINTS

- The exercise of power within an organization is necessary in order that its goals are achieved.
- An organization will never be very rationally driven; the personal objectives and interests of its members may oppose those of the organization and create inefficiencies and conflicts. Power must therefore be exercised in order to avoid a situation of personal interests prevailing over organizational goals.
- There are five types of power: reward power, coercive power, legitimate power, expert power and charismatic power.
- An organization is influenced by external and internal powers.
- Leaders must understand how power is distributed and try to enhance their own power; the most effective way of doing so is to increase others dependency on them.
- Conflicts between power holders are normal. The most frequent can occur between political authorities and operational leaders (over objectives, due to egos); splits between headquarters and divisions or subsidiaries are another area of frequent tension.
- Good leaders know to identify and concentrate on their own contribution; they know how to surround themselves with people who compensate for their shortcomings and they know how to adapt their style of leadership to situations and personnel.
- Situational leadership theory offers an interesting tool for leaders in managing flexibility according to the maturity of their personnel.
- Situational leadership can also be useful for adapting the leadership style to a new policy.

Chapter 5

Command in Operations

Objective

Chapter 4 discusses power and leadership in general terms. This Chapter focuses on the issue of command in operation: firstly, the difference between command and management; secondly, the particularities of command in police organizations compared to within the military; and thirdly, the role of staff, the rhythm of command through briefings, and the specific role of the staff officer in charge of intelligence.
In this chapter, many references are made to the military; this is due to the fact that the military model has often times been adapted by most police forces in the conduct of operations involving the deployment of the structures described in Chapter 3.

5.1 COMMAND

5.1.1 Definition

In defining command, this book shall refer to Martin van Creveld.[1] Although he addressed the issue of command in a military environment, his comments are applicable in the police context and assist in illuminating the relationship between management and command. Van Creveld wrote, "Command may be defined as a function that has to be exercised, more or less continuously, if the army is to exist and to operate… the responsibilities of command… are commonly divided into two parts. First, command must arrange and coordinate everything an army needs to exist – its food supply, its sanitary service, its system of military justice, and so on. Second, command enables the army to carry out its proper mission…; to this part of command belongs, for example, the gathering of intelligence and the planning of operations." A few lines later, van Creveld noted that, "another and equally useful way of looking at command is to ask not what its responsibilities are but

1 Martin van Creveld, *Command in War*, Harvard University Press, Cambridge, Massachusetts, 1985, Introduction.

what it does…There is, in the first place, the gathering of information…objectives must be laid down and alternative methods for attaining them worked out. A decision must be made; detailed planning must be conducted. Orders must be drafted and transmitted… Execution must be monitored by means of a feedback system, at which point the process repeats itself."

5.1.2 Differences between command in a military and police environment

This passage was cited not only because it is reminiscent of the rational decision-making process described in Chapter 2, but also because it allows one to observe the differences between command and management, while underlining some of the differences between military and police. Van Creveld does not speak of management, but what he calls the "first part" of the responsibilities of command could also be understood as the managerial activities necessary to prepare the army to accomplish its operations.

Management can be defined as "managing an organization, managers, workers and work."[2] Drucker also provides this definition: "defining the goals and mission of the organization, executing the tasks efficiently, managing the social impacts and the social responsibilities." Command in the latter definition is fully integrated into the management function, as it is one of the conditions of the exercise of this role. However (and this is a key difference between military and police activities), the same separation between preparation and execution does not exist in police tasks as is the case in military operations. The activities of the former are ongoing, while those of the latter are generally limited in time (even if it is over a long period). Fighting crime, and maintaining order and security are fixed activities; waging war will have an end, through either total victory, an armistice or a withdrawal. Armies will then return to their peacetime activities and to the preparation phase. Police forces will always be required to conduct managerial and operational activities in parallel. Command will be a part of their activities, but will remain fully integrated within the existing managerial process. In times of military operations, everything will be subordinated to the requirements of command and all efforts will be concentrated on the achievement of the objectives set; in policing, there will naturally also be objectives, but the variety and diversity of the problems will always prevent police from reaching clear-cut victories. Tactical and operational successes will be achieved, but total victory remains impossible. This situation also explains why there are differences in the profiles of commanders. A peacetime commander or chief of the army may have to be replaced during war, as the qualities needed in both situations are different. A police chief must be an excellent administrator first; he or she must naturally aim to be surrounded by operational talents, but his or her main responsibilities will concern the effective operation of the police force and he or she rarely will – at least in a large police force – exercise direct operational command. Table 5.1 presents a chart that underscores some of the principal differences between military and police.

---

2 Peter Drucker, *Management, Tasks, Responsibilities, Practices*, Management Editions (Europe) Dirmas S.A., Geneva, 1974, 1 and 39ss.

Table 5.1 Differences between military and police activites.

Aspect	Military	Police
Types of problems faced	*Preparation*: to develop an efficient organization prepared to conduct war. *Operations*: to achieve the military objectives set by politicians.	To develop efficiency in the organization, methods, and use of resources to handle a diversity of problems.
Nature of problems	Normally, time bound.[3]	Recurrent.
Opponents	Normally clearly identifiable.	Diverse and often disguised.
Concentration	Possible at every level: strategic, operational, and tactical.	Possible at operational and tactical levels for operations limited in time. At the strategic level, goals are more in relation to policies and change management than operations.
Forms of deployment	Chiefly in large and organized formations.	Interventions by individual police officers or by small units are the rule; large deployments – generally of ad hoc formation – are the exception.

The term command will therefore be used in the policing context only for operations; the term management will be reserved for the general activities of an organization. Military command refers to all activities concerning the preparation and operations of the military.

5.1.3 Issues concerning operational command

The execution of an operation, even if perfectly planned, is always going to present a certain number of surprises or conditions that had not been taken into account. The management of information as previously stressed is a key component notably in the capacity to decide on an appropriate level or quantity. The existence of a reserve (see Chapter 2, section on *Tactics and Deployment*) is similarly crucial in order that the chief of operations can maintain a freedom of maneuver. A further issue must be dealt with by the Chief of Operations (CoO): the degree of autonomy left to his or her subordinates. The chief's plan of action must thus be understood by his or her subordinates in order for them to abide by his or her responsibilities without having to wait for orders (as long as the basic idea of the maneuver is respected).

The time pressures of the problem faced should be reflected by the style of command. In the planning phase of operations, it is strongly recommended that commanders employ a participative approach. During implementation, the command style will become more directive.

[3] Although modern conflicts are taking on new forms, they occur in more or less clearly span of time, identifiable by a beginning and an end marked by an armistice or peace treaty.

5.1.4 Levels of decision-making

Different levels of decision-making should be considered according to the importance of the case in question, as described in Chapter 1. Here it is useful to consider who are the police stakeholders, as they may vary according to case. Likewise, emphasis should surround the core missions at hand, rather than on organizational details. Table 5.2 summarizes these measures:

Table 5.2 Core missions at hand.

Type of case	Actors involved	Mission(s)
Normal: the police officer or patrol(s) can handle the case.	Police officer/ patrol.	To act (to rescue, secure, protect evidence); To structure (if necessary to task personnel on the spot); To inform.
Specific: the police officer or patrol does not have the resources needed to handle the case.	Command center.	Send reinforcements; Search for complementary information; Recommend measures; Inform higher levels.
	Duty officer.	Check measures taken; Take complementary measures if necessary; Go to the location if necessary to take charge.
Crisis: the size of the case requires significant resources; the coordination of several stakeholders may be required; the duration of the intervention is long.	Specific crisis management staff.	To take over the case.

5.2 THE ORGANIZATION OF COMMAND

5.2.1 The chain of command

The chain of command is the formal line of authority, communication, and responsibility within an organization; it ensures clear assignment of duties and responsibilities.

5.2.2 Direction and line

The previous chapter introduced the different levels of decision-making developed by Mintzberg. These can be reproduced in the organization of command (Table 5.3). The primary line elements of a police department are uniformed patrol, criminal investigation, and traffic.

Case Study: Swiss and French command structures during the G8 Meeting of 2003 (Evian)

In 2003, the G8 meeting took place in the French town of Evian. Participants landed in Geneva and were driven to Lausanne, where most were accommodated and later transported by boat to Evian for the meetings. Demonstrations were held in Lausanne and Geneva but not in France. Close cooperation was established between both countries in order to ensure the security of the summit. The political and organizational differences between both countries were reflected in their respective chains of command.

On the French side, all the security forces reported to the *Préfet* who was directly responsible to the Presidency. This conferred upon him a large decision-making power, and he could act very rapidly. On the Swiss side, complex structures – political and operational – were established in order to respect the different jurisdictions, as three cantons and the federal state were involved. The decision-making process was therefore much longer and more complicated than the French one.

Table 5.3 Organization of command.

Level	Organization	Command
Direction	Chief	Elaborates the vision, defines priorities and objectives; allocates resources.
Management	Senior officers	Develop operational plans; May be tasked to lead large operations;
	Middle managers	Responsible for implementing operational plans; Formulate objectives and action plans in their areas of responsibility;
	Sergeants	Lead and supervise.
Manpower	Police officers	Execute plans.
Support personnel and analysts	Civilians	Facilitate and support operations.

5.2.3 Front line staff

As organizations grow, their functions become more specialized. Staff components are thus added to the line structure. Staff personnel will provide expert advice to line units (scientific, legal, and organizational); they will accomplish specific tasks (analysis, communication, etc); and they will contribute to achieving department-wide conformity in activities that affect the entire organization.

In the management of large operations, one will find the same co-habitation between line and staff personnel. The staff personnel will be responsible for the following tasks:

- Help to prepare decisions to be made by the chief; *the division of tasks* is very clear, with the chief making decisions and staffers providing him with elements

to aid in the decision-making. When a decision has been made, staff will contribute to its execution. Staff will also work on identifying new risks that could emerge and develop options related to measures to be proposed to the chief.
- Contribute to the execution of tasks.
- Assist in the coordination of forces deployed.
- Oversee the execution of orders given and provide feedback.
- Identify potential developments and prepare counter measures.

The following represent key steps contributing to the success of staff support:
- A clear organizational chart has been designed and produced.
- Job descriptions have been prepared for each staffer.
- A method formalizing the decision-making process has been developed and disseminated.
- Procedures define how work should be carried out (organization of shifts for instance).
- Information is effectively disseminated in a timely fashion.
- Working conditions are adequate (such as rooms and materials).

5.2.4 Timing considerations linked to staff work

Three different time dimensions must be simultaneously addressed: the past must be analyzed in order to understand and identify the root causes of an incident; the measures subsequently undertaken must be monitored and assessed; and future developments must be considered, including an assessment of possible consequences and risks.

Staff composition

The structures indicated in Chapter 3 serve as basis for different situations. Staff composition will, however, have to take into account the specificities of each separate situation. The positions outlined below will therefore need to be completed or modified where necessary:
- The operation chief makes decisions and interacts with relevant political authorities or other stakeholders.
- The deputy-chief may coordinate the work of the staff.
- The secretary prepares meetings and takes minutes.
- Officers from the uniformed police, criminal police and special units deal with tactical issues.
- The intelligence officer acts according to the tasks described in section 5.3 (briefings).
- Support specialists assume responsibility for transportation, telecommunications, equipment, finances and accommodations.
- The media officer handles external communications.
- Additional specialists may be needed according to the situation (such as legal advisers, psychologists, etc).

Within this organization, *ad hoc* cells (permanent or temporary) may be set up to address specific problems. The chief may decide to task some of the staffers to prepare a set of recommendations for how to address a particular issue. External personnel can be brought in where necessary.

5.2.5 Expected qualities of staff members

Individually members of staff must have the necessary skills to accomplish their specific tasks. As members of a group, they must be team players, be disciplined, and have an analytical mind capable of facilitating the decision-making of the chief of operation (involves the capacity to understand the chief's ideas).

Case Study: Churchill's organization during World War II[4]

The problems faced by Churchill in directing Great Britain's war efforts were many and complex; Churchill set up an impressive machinery to manage them. The main features were:

- Structures: the war cabinet was Britain's principal policy-making body, and regularly attended by the chiefs of staff as well as by 8 civilian members; some 400 committees and sub-committees branched from it.
- Information: a Defence Registry logged every incoming signal from commanders in the field; it provided the basis for the excellent exchange of information that prevailed between Downing Street and the service ministries.
- Coordination: the chiefs of staff met every day except Sunday at 10:30: they attended the war cabinet meetings every Monday evening. Major-general Ismay was appointed to ensure good liaison between the Prime Minister and the service chiefs.
- Work methods: staff discipline was high; minutes were typed and circulated within an hour or two of meetings taking place, even after midnight; private secretaries worked in shifts through the day and much of the night.

Churchill was also aware of his own weaknesses and established rules to keep them in check. He was impulsive and intemperate in judging people and plans, and liked to voice many new ideas (whether realistic or not) that could confound his interlocutors. His close circle knew that he often spoke merely as a means of helping himself to formulate ideas. To reduce the effects of his prolixity, he established a clear rule that nothing he said was intended as a basis for action, unless subsequently confirmed in writing.

The most notable aspects of the machine for the direction of Britain's war effort was that it was better ordered than that of any other, including those of Germany and later the US.

4 Max Hastings, *Finest Years, Churchill as a Warlord 1940-45*, Harper Press, London, 2009, pp. 140ff.

5.3 BRIEFINGS

The decision-making process represents a primary factor of success in the work of staff. The process was examined in a generic sense in Chapter 2; the following section will discuss how staff applies the rational model. It should be re-emphasized that this book presents an ideal model that may, in reality, be subject to constraints. At the same time, the method constitutes a sound basis for the work of staff and operational problem solving. Considering issues of procedure in detail, meetings and briefings provide the organizational framework within which measures are decided, implemented, monitored and assessed. Meetings and briefings represent a process with a specific rhythm, though involving variations and differing roles for key members of staff. While the military model is the basis for these developments, many police forces have adapted it to their own needs and use it with notable efficiency.

The "rhythm of conduct" refers to the logical order in which issues are handled by staff. New incidents that occur in the midst of the process will have to be integrated into assessments and decisions being made, though handled via an application of the same method. Missions, options and new developments should be assessed in the following manner:

When a mission is received or a problem has to be addressed, staff (or the chief of operation) takes a certain number of immediate measures in order to gain time to assess the situation in depth:

- Analyze the mission, situation or event.
- Determine urgent measures to be taken (to alarm, mobilize, organize, inform, and search for information).
- Establish a time frame.
- Consider organizational issues (*e.g.* setting up a specific task force).

The next phase is devoted to a detailed estimate of the situation and the subsequent development of possible options. Orders are formulated and decided on this basis.

In the third phase of the process, the orders are disseminated.

Lastly, implementation of the measures is monitored (including corrective adjustments introduced if necessary) and the results assessed.

Throughout the process, potential developments and their consequences are observed in order to prepare counter-measures.

5.3.1 Types of briefings

In theory, there are three types of briefings, each corresponding to one of the three phases described above (the monitoring of measures in process should always form part of the agenda of meetings). In practice, however, the strict division between briefings will be rare, due to time pressures, a lack of information, and the simultaneity of different events. Therefore, some of the seemingly discrete stages of the model will have to be omitted or merged. Nevertheless, it is still useful to describe how staff meetings should be organized even at a 'theoretical' level.

- The *situation briefing* corresponds to phase one and addresses the issues indicated, including the points outlined in Table 5.4:
- The *assessment briefing* corresponds to phase two, and there can be two types of situation that constitute the focus of the meeting:
- The beginning of an event: for instance, where security measures are being planned for a future event, the identification of problems will have been completed in the first meeting, while the second will proceed with an assessment of the situation (see Chapter 4).
- The execution of an operation: where the purpose of the meeting is to assess progress, identify possible developments and prepare possible measures. This kind of meeting can follow the agenda shown in Table 5.5.
- The *operations order briefing*_corresponds to phase three (Table 5.6).

Table 5.4 The situation briefing.

What	Who	Remarks
Purpose of the meeting	Operation chief	
Situation	Intelligence officer	
First analysis of the mission, situation or event	Operation chief	
Urgent measures	Staff	One of the key measures involves defining the information needed.
Time frame	Operation chief	Establishing a calendar of operations by counting backwards from the final deadline.
Organizational measures	Operation chief	
Questions		
Synthesis	Operation chief	
Next meeting and specific deadlines	Operation chief	

Table 5.5 Meeting agenda for operation execution.

What	Who	Remarks
Purpose of the meeting	Operation chief	
Situation	Intelligence officer	
Analysis of the situation in the different sectors	Chief of operation and specialists	Measures previously ordered are monitored and assessed.
Possible developments	Sector specialists	Identification of possible developments and potential consequences (including likelihoods).
Measures • Urgent measures • Others	Staff	One of the key measures involves defining the information needed.
Time frame	Operation chief	Establishing a calendar of operations by counting backwards from the final deadline.
Organizational measures	Operation chief	
Questions	Staff	
Synthesis	Operation chief	
Next meeting and specific deadlines	Operation chief	

Table 5.6 The agenda for an operations order briefing.

What	Who	Remarks
Purpose of the meeting	Operation chief	
Situation	Intelligence officer	
Mission	Operation chief	
Intention	Operation chief	Summary of the plan to be used in accomplishing the mission.
Execution	Operation chief	Explanation of the mission to relevant subordinates (direct subordinates will be present).
Administration and equipment	Specific managers	
Control and communications	Specific managers	
Locations	Operation chief	
Questions		
Synthesis	Operation chief	
Next meeting and specific deadlines	Operation chief	

The efficiency of the briefings depends on different factors:

- Prior preparation (the agenda).
- The conditions of the meeting (room, material).
- The management of the meeting by the CoO.
- The quality of information provided.
- The contribution of staff members (what they provide and how).
- The clarity of conclusions reached (what, who, when).

Case study: The Mission Analysis Briefing of NATO[2] (a format outline)

1. Introduction
    - Aims of briefing
    - Outline
2. Review of situation
   General
    - Political, military, and historical background
    - Current situation
    - Opposing force situation
    - Intent
    - Capabilities
    - Assessment (brief description of likely actions)
    - Higher Commander's direction and guidance
    - Mission
    - Intent
    - Objectives
    - End state and criteria for success
    - Assumptions
    - Limitations
    - Tasks assigned to your command
    - Tasks implied for your command
3. Operation design
   Key factors and obstructions (only those that impact mission analysis)
   Key strengths and weaknesses
    - Opposing force
    - Own force
    - Relevant other forces (as necessary)

   Centers of gravity
    - Strategic
    - Operational
    - End state and criteria of success
4. Proposed mission
5. Initial force estimate
    - Forces available for planning purposes (if given)
    - Estimate of forces required

6. Proposed planning guidance
    - Proposed initial intent
    - Direction to planning staff
    Direction to subordinate commands
7. Conclusion
8. Approvals, seek approval of:
9. Operation design
10. Mission
11. Proposed planning guidance

5.3.2 The intelligence officer

Annex 3.2 outlined the tasks of the chief of an organizational intelligence unit, which can summarize as follows:

- Manage the cell.
- Manage the collection of information (including issuing orders for that purpose).
- Analyze and utilize information.
- Disseminate information.

Table 5.7 Requirements faced by the intelligence officer.

What	How
Propose urgent measures linked to his or her area of responsibility.	Collect information about the adversary and terrain.
Present an overview of the situation.	General situation and developments since last meeting. Adversary: who, how many, equipment, where, what are they doing? Organizational resources (as above).
Prepare an estimate of the situation.	Terrain: description with identification of the strengths and weaknesses for the opponents and possible threats. Weather. Adversary's options: what can they do, when, how, for what purpose? Most dangerous threat (planning phase, most likely risk (execution phase).
Prepare orders.	Directions for information collection. Proposals for certain points of the operation order issued by the chief of operation (situation, secrecy rules).
Monitor.	Oversight of the order(s) issued.
Manage permanent operational activities.	Intelligence gathering, fact finding Information assessment, picture building analysis, recommendations, dissemination.
Manage post-operation follow-up.	Provide information in his or her area of responsibility useful for assessing the course and results of the operation and establishing lessons learned.

As an operational staff officer, the intelligence officer fulfills two types of tasks: manage the intelligence unit and provide information and analysis to the chief of operation to support the decision-making process. Thus, he or she plays a prominent role.

As a member of the operational staff, the intelligence officer will, depending on different meetings, be faced with a number of requirements (Table 5.7).

5.4 SUMMARY AND KEY POINTS

- "Managing" in an operational context, means commanding. Police forces have generally adapted the military model of command for conducting operations involving the deployment of the operational structures described in Chapter 3.
- Decisions are implemented within these specific operational structures through a process of decision-making structured through different reports.
- The chain of command – the formal line of authority, communication and responsibility during operations – establishes the relationships between direction, line and staff; it must remain clear and simple.
- Members of staff provide specialized information concerning their specific areas of activity to the chief, and contribute to the coordination and control of forces deployed.
- Three levels of time are addressed simultaneously: the analysis of the past in order to understand the roots of the problem; an analysis of the present in order to monitor and correct the ongoing actions, and an analysis of the future in order to prepare countermeasures to possible evolutions of the situation.
- Information plays a vital part in each of these phases and therefore the intelligence officer is a key member of the staff.

Chapter 6

Training

Objective

Most police organizations have developed broad competencies for delivering education and training programs. This Chapter will address some general issues, and discuss certain hesitations pertaining to police training. Mostly importantly, the chapter will propose how to best develop a training program or seminars around decision-making frameworks described in this book.

6.1 INTRODUCTION

Police training has developed rapidly over the last decade. Until recently, experience was seen as a unique means for transmitting knowledge. Police schools were essentially dedicated to the socialization of recruits into the profession. For example, discipline, the learning of legal procedures, the use of guns and fighting techniques, as well as the practice of sports were considered as central training elements. Some noticeable exceptions exist, such as programs devised by pioneers at the beginning of the 20th century,[1] which remain applicable in contemporary contexts. Only recently, however, has the complexity of policing been generally accepted, as well as the value of knowledge originating outside the police world.

Without fully rejecting the socializing role of training, many other kinds of courses have developed under the umbrella of police schools and "academies." Universities also offer relevant education, either fully or partially dedicated to police officers. Such programs provide an opportunity to consider practical applications of relevant theoretical approaches (*e.g.* criminological or forensic science), as well as to benefit from findings grounded in empirical research.

When devising the structure of a police-training program, the search for the right balance between theory and practice, generality and specialization, or methods and techniques, is an ongoing challenge. At the same time, the decision-making methods explained throughout this book remain consistently applicable. As shown in Table 2.57, it was suggested that differentiated training programs address

1 See for instance August Vollmer and Albert Schneider, "The School for Police as Planned at Berkeley", *Journal of the American Institute of Criminal Law and Criminology*, 6 (1917), pp. 877-898.

specific actors (*e.g.* special units, leaders, investigators or crisis management cells). Another perspective exists, however, that may impact more profoundly upon how to conceive of training activities. Indeed, it has been argued that decision-making ensures some unity across police work. That is, the viewpoint provided by decision-making forms a common thread upon which to base a coherent training framework.

6.2 WHO: TRAINEES AND INSTRUCTORS

Searching for the best combination of instructors and trainees is a principal aspect of training.

6.2.1 The trainees

Police personnel are concerned by key concepts almost independent of an individual's position in the organizational structure: training examples and case studies have been borrowed from criminal investigation, crisis management, intelligence, and many other activities carried out across a force. Moreover, the methods presented can benefit users throughout their career. Indeed, an appreciation of the difficulties and responsibilities underpinning decision-making can assist young officers in orienting their careers around "real" capacities and expectations. Senior officers may also benefit from such material, as a basis for structuring their acquired experience. Managers too can gain from new knowledge about how to integrate structured decision-making processes in an organizational setting. Such training is an opportunity to bring together people with very different roles, missions and tasks, backgrounds and seniority. It may create cross-fertilizations, counter the development of sub-cultures, and initiate networks of confidence.

There are, nevertheless, some difficulties and a balance needs to be found when assembling a group of trainees. Direct hierarchical relationships between the participants may play a negative role – in particular, by hindering free expression, which is crucial for the exchange of experience. Additionally, it tends to be an advantage if trainees are selected on a voluntary basis. The dynamic of a group depends significantly upon individual motivations.

6.2.2 The instructors

Developing a training course requires some reflection about the profile of instructors versus participants, a combination that is key to successful learning outcomes. Practitioners generally demand that instructors demonstrate a high level of credibility by fully integrating into their specific communities (that is, possessing shared experience). However, the debate between choosing a model featuring professional instructors, or the intensive involvement of practitioners, is still open for debate. Viewed simplistically, professional trainers are often perceived to have expertise in teaching, but also to have lost contact with the "real world" (and tends to incur high costs), while practitioners are seen to tell a series of impressive stories that are close to daily work, but generally anecdotal. This solution also disguises the costs of training, which is subsumed into the regular salary of individuals who become occasional instructors. No definitive solution exists aside from searching for a good balance, and choosing the right people according to well-defined objectives (*e.g.* communication skills, ability to moderate a debate, credibility).

The intervention of civilian experts or other specialists may also add value if they are well recognized as such. The priority, however, should be selected based on an individual's ability to communicate their expertise. In order to offer the opportunity to trainees to discuss possible conceptions of their practice, or confront these points of views with existing theories, it may be appropriate to include academics as instructors; this issue will be discussed further below.

6.3 WHERE: LOCATION OF THE TRAINING ACTIVITY

Where a training course takes place is of high importance – the quiet atmosphere of a lecture room contrasts markedly with the daily pressure of the actual job. There are, however, advantages and disadvantages:

- Disadvantages:
  - Whatever the quality of a training program, the gap between exercises and actual practice remains important.
  - Participants used to daily intensive outdoor activities may not be accustomed to the inactiveness of a lecture-style class environment.
- Advantages:
  - Participants are plunged into a completely different context that insulates them from the pressure of their daily work.

Seminars organized *extra-muros* across several days offers the advantages of possibly creating an atmosphere that increases confidence between the participants and instructors, and favors exchanges of experience. Such an approach may also level out hierarchical relations between participants. At the same time, the costs may nonetheless deter organizers.

6.4 WHAT: LEARNING OUTCOMES

Some practitioners may show little enthusiasm for conceptualizing and integrating new methodological or intellectual frameworks. This is particularly evident for decision-making, where the overall aim centers on shaping attitudes and the use of general processes to help participants deal in a flexible way with the variety of problems they face in their professional lives. This kind of training is not about the acquisition of a series of basic techniques reinforcing an actual practice. The balance between theory (knowledge) and practice (know-how) should be emphasized in program design. Fundamental aspects and difficulties of a task should not be hidden behind a demagogical and reductionist attitude by overly simplifying complex problems in order to secure positive evaluations.

In this context, one may anticipate problems related to expectations by specifying carefully and at the onset, anticipated learning outcomes. Several models for this purpose exist; however, they all concentrate on what the participants should have learned or should be able to do after having followed the program.

The models can be stated in the following way:

- Reconstruct the operational chain, describing key principles underlying each step.
- Reference the different methods proposed.
- List difficulties, limits and traps related to the application of these methods.
- Illustrate different aspects through personal experiences.
- Apply the methods.
- Examine each step in the context of the organization of participants, including how to adapt the general process for this purpose.
- Draw out the connections in order to ensure a level of coherence across the training activity as a whole.
- Advocate to defend the application of the covered methods in participants' organizations.

6.5 HOW: GENERAL RECOMMENDATIONS FOR TRAINING DELIVERY

The training style is dependant from the objectives (learning outcomes). One might chose for instance to confront viewpoints, stimulate conceptualization from personal experiences, present existing theories and their scope of application, promote learning by doing, or implement a "trial and error" learning strategy. There is no definitive answer whether one training style is better than the other. However, participants bring with them their experience which cannot be immediately replaced by new representations. Mental models built from the vividness of real cases are very resistant to novelties. Systematically taking into account this background in the learning process is a basic principle. In this perspective, looking for active participation is always preferable when practitioners are engaged in a training program. The dynamic of the course is generally better when the types of activities are varied. Lectures, working groups, think tanks, personal work, or exchanges of specific experiences in plenary sessions may be alternated.

6.5.1 Theory versus practice: a never ending story

One of the main problems in designing training activities is connected to the need to carefully integrate concepts and methods with technical skills. Decision-making is not a technique to be aimlessly applied, but requires a deep understanding of concepts that underlies the way the presented tools can be used. As an anticipatory measure, it may be useful to remind participants that their daily practice offers little space or time to conceptualize their activities, as people tend to be habitually and emotionally engaged in their daily activities. Seminars and courses are unique places where they can do so. Whatever the strength of this argument and the manner in which learning outcomes and objectives are presented, programs are still often assessed by participants as "too theoretical."

In particular, when researchers and academics are involved, the misunderstanding between *a priori* opposing views of "theory" and "practice" can occasionally be amplified, and may exacerbate rejection of the covered content by the audience. More often than not, this response can occur independently of the value of the content or the quality of instruction. Rather, it appears based mainly on misunderstanding and occasionally also on an aversion to intellectually demanding activity. Specific solutions can be found.

Case Study: a university professor engaging with practitioners

A professor of criminology from a University was convinced that certain types of criminality were decreasing. His scientific studies, based on solid statistics, showed the trend very clearly, but the message was not accepted when conveyed to practitioners that dealt with this form of criminality. Indeed, the professor occasionally even faced whistling at the end of his lectures. Both sides believed their own view to be correct.

As a result, the professor decided to change his approach by considering the point of view of the audience. Before presenting his statistical material, he started his lecture by explaining the traditional bias of human beings confronted by a reality that stimulates certain emotions: "information that is vivid, concrete, and personal has a greater impact on our thinking than pallid, abstract information that may actually have substantially greater value as evidence."[2]

In this way, the practitioners' activity was both recognized and considered, helping to concentrate attention on the objective figure relating the drop in criminality. The professor was subsequently applauded at the end of his presentation.

6.5.2 Planning/Progression

A plan of an entire seminar can be oriented towards learning outcomes by specifying intermediate steps. An instructor can use the myriad of appropriate frameworks developed by police schools, as well as targeted support for delivering presentations, using information technologies or organizing group work. Parts of a program may be developed through e-learning, which involves many facets. E-learning may be designed to collect results achieved by students when they have completed a computer-assisted exercise. Indeed, automatic collection of results provides some empirical basis for instructors to point to repetitive mistakes.

2 Richards J. Heuer, *Psychology of Intelligence Analysis*, Novinka, New York, 2006, p. 126.

There have also been promising developments in using e-learning for distance courses. Removing personnel from daily operational activities is generally seen as an organizational sacrifice. Similarly, travel distance can be an impediment when, for example, a training activity concerns several countries. At the same time, this form of training course must still find a place within police training programs. Though e-learning is proliferating as a training technique, an effective balance must still be struck between on site and distance courses, especially when considering the social role of training.

6.5.3 Exercises

A training program necessarily includes exercises that simulate real situations. Experience has demonstrated that scenarios strongly based on actual cases are preferable– a coherent scenario of a sufficient complexity is almost impossible to build from scratch. The complexity of scenarios must, however, fit with time constraints. Exercises are tested carefully in advance.

A single scenario can be used to draw together the various elements of the operational chain. For instance, a vehicle theft scenario could be built around the following progressive sequence (Table 6.1):

Table 6.1 Vehicle theft scenario.

Given to trainees	Step of the exercice	Pedagogical aims
The perpetrator of a vehicle theft has been arrested and material has been found in his vehicle (give the inventory).	Choose actions to be taken according to the description of the situation.	Understand how application of the operational chain can help structure the investigative approach (Chapter 2, *Operational Chain*).
Other similar cases occurred in the region during the same period.	Evaluate the problem.	Learn how to ask for information: define requirements and needs (Chapter 2, *Decisions*)
Assume the position of the analyst.	Produce profiles (who, what, where, when, how, with what, why)/ maps/chronologies.	Understand the intelligence cycle (Chapter 2, *Information*)
Organize a decision-making meeting for responding to the problem analyzed.	Prepare the room, distribute the roles, carry out the meeting and formulate an order.	Learn how to prepare a briefing (Chapter 3, *Structure*), develop possible actions and tactics, and deploy resources (Chapter 2).

The exercise duration, boundaries and organization are carefully explained to the participants, as well as the aims for each step (in line with the theoretical backdrop). When organizing group work, methods such as brainstorming (Chapter 2, *Decisions*) can be applied in several steps. Possibilities to discuss issues of command and leaderships are also manifold (Chapters 4 and 5). When completed, each step can be further considered in plenary, by letting participants present what they have accomplished (otherwise, a lack of recognition could result in frustration or detachment).

6.5.4 Miscellaneous

For the different reasons mentioned, courses devised for practitioners are as interactive as possible. The role of trainers as moderator is important in order to maintain control and allow full participation. Carefully organizing rooms and working groups is a key factor of success in this perspective. Trainers should look out for situations where a limited number of individuals assume leadership, hampering the active participation of others. Any technologies such as mobile phones and even computers that may divert attention can be removed during the course, an approach that generally results in greater concentration and more interaction during the sessions.

6.5.5 Evaluation

Did a program achieve its objective? Was the structure adequate? How good was the performance of the instructors? Evaluation is a key aspect for adapting programs on the basis of lessons learned and there are several approaches one can take. Generally, in order to answer some of these questions, trainees are tested. However, participants are also practitioners who may have had significant prior experience in multiple forms of evaluation and assessment. There is a tendency today to exaggerate the intensity of controls and assessments that may imply some doubt concerning the professionalism of participants. Some tests may be understood more as a pretext for meeting formal quality requirements (rather than providing real added value), or as a pretext for surveillance and the exercise of power. Tests may place the participants under unnecessary pressure. Good selection of participants and the creation of a confidence-boosting and positive atmosphere that encourages participation and engagement during the course can be more efficient than systematic testing.

An advantage of testing, however, is to inform instructors about how concepts have been understood, and if principal messages have been adequately communicated. Weaknesses in the design of a program or in the chosen pedagogy may be detected in this way. Yet, collecting material provided during plenary sessions and exercises generally provides sufficient empirical material for evaluating how fundamental messages have been received and for preparing collective or personal feed-back to trainees. Another way of evaluating a training program is through a questionnaire to be completed by participants upon conclusion of activities. A questionnaire would generally address many aspects of a course and be oriented around the stated objectives. Typical questionnaires can be found easily in any professional environment where courses are delivered. Trainees answer anonymously questions such as:

- Did the course fulfill your expectations?
- Were the objectives well-defined?

- Was the course well-structured?
- Was the course suited to the level of participants?
- Was the course well-presented (pedagogy, pedagogical support, technology)?
- What were the strengths and weaknesses of the course? What could be improved?

At the same time, there are many limitations in interpreting such questionnaires. Indeed, results may be biased as experience is the result of individual attitudes acquired on a case-by-case basis. Thus, good evaluations of a program by participants may mean that the course has conveyed knowledge that has reinforced an actual practice rather than delivered a fundamental message. Conversely, poor evaluations may also reflect some resistance to the notion that complex tasks addressed in daily practice require demanding effort in order to change attitudes.

6.6 SUMMARY AND KEY POINTS

- Training concerns everybody and may be organized in order to favor cross-fertilization and the realization that decision-making gives some unity to the police work.
- Decision-making training should not aim to reinforce an actual practice but rather a change of attitude.
- Theory and practice should be balanced, even if trainees ask for more "practice."
- E-learning can be a valuable tool for supporting part of a training program.
- Good exercises require extensive preparation.
- Though careful evaluation of participants and programs is suggested, this may have undesirable effects.

Chapter 7

Conclusions

The decision-making process has been chosen as the main topic of this book because it is one of the key mechanisms in the management of a police organization and in the conduct of its operations. For decisions to be effective, they must provide adequate answers to identified problems, mobilize necessary means and deploy them rationally. Decisions are not made in a vacuum, they are made as part of a process that is embedded within given structures, executed by human beings (with all the constraints implied), subject to limitations linked to available resources, information, time, and regulations, and vulnerable to internal and external influences. That is why specific chapters of the book have been devoted to the environment of the decision-process: the structures, the leadership, the command mechanisms, and the training.

Many references have been made to the military and the business world as they have largely influenced the ways police forces manage operational and administrative problems. One should not however consider that the methods and leadership styles employed by the military and private companies could be transferred as such to police environments. Situations, constraints, legal frameworks, and human and social dynamics are different; these methods have to be adapted to the specific requirements of the police.

The principal theme that runs through the entire book is that the use of a rational method aids in bringing a level of order to what would otherwise be chaos and that human factors, even if often unpredictable, can be made more reliable if decisions are made according to a clear methodology and with appropriate leadership. The importance of coherence has been stressed throughout the book: coherence between the elements of a system; coherence within the approach to a problem; coherence between ends, ways, and means; and coherence between situations and corresponding structures, leadership and command styles and processes.

The book has devoted much attention to intelligence because, at each specific stage of the process, information plays a prominent role in providing elements for making a decision, highlighting options, and monitoring and assessing subsequent actions. Police mainly used information to reconstruct the past in judiciary investigations and for monitoring ongoing actions. New and promising developments are underway to help police to better target its efforts and deploy its resources more efficiently as well as to better identify the trends and the causes, which shape the evolution of crime in order to develop appropriate solutions.

The book underscores, in the decision phase, not only methodological issues, but also the interference of human factors, which can either distort or support the decision-making process. Hidden traps (prejudices, biases, personal interests), or poor leadership can strongly influence – whether positively or negatively – outcomes. There has been a particular emphasis on the necessity of achieving coherence between a decision, the deployment of resources and the use of tactics. Success in deciding and implementing is related to the general framework in which these processes take place. Therefore, responsible managers must thoroughly analyze their organizations and its mechanisms in order to make necessary changes.

For this reason, it is important in conclusion to highlight the issue of change management. It is widely considered that technical changes are easy; if a new weapon is introduced, police officers will soon be able to use it. Contrastingly, changes concerning procedures are more difficult and take more time because work habits have to be modified. The most difficult changes concern the structures and policies; they affect the environment in which members of an organization work in and the manner in which they accomplish their tasks. Such changes can be disconcerting, as they create much uncertainty as to an individual's place or even future in an organization. The time needed to implement such changes is generally lengthy – months or even years. It is for this reason that a chapter has been written on training, as this dimension is a primary component in all reforms.

It has been our attempt in this book to develop an integrated approach in placing the decision-making process within a larger context. This approach is in fact very demanding because it requires police organizations to address many issues in order to establish an efficient decision-making process but we are convinced that these changes can tremendously improve policing.

Bibliography

Books and books sections

Bazerman, M. H. and Chugh, D., "Decisions without Blinders," *Harvard Business Review*, Special Issue January 2006.

Blake, R. and Mouton Jan S., *The New Managerial Grid: Strategic new insights into a proven system for increasing organization productivity and individual effectiveness, plus a revealing examination of how your managerial style can affect your mental and physical health*, Gulf, Houston, 1978.

Cohen, E. A., *Supreme Command*, Anchor Books, New York, 2003.

Cusson, M., "Questions de stratégie pour la police", in: Cusson, M. B. Dupont and F. Lemieux (eds), *Traité de sécurité intérieure*, Presses polytechniques et universitaires romandes, Lausanne, 2008.

Cusson, M., Dupont, B. and Lemieux, F., *Traité de sécurité intérieure*, Presses polytechniques et universitaires romandes, Lausanne, 2008.

Dolan, S. and Lamoureux, G., *Initiation à la psychologie du travail*, Gaëtan Morin, Boucherville, 1990.

Drucker, P., "Une decision efficace", in: Collection Harvard Business School Press, *La prise de decision*, Editions d'organisation, Paris, 2002.

Drucker, P., *Management, Tasks, Responsibilities, Practices*, Management Editions (Europe) Dirmas S.A., Geneva, 1974.

French, J. and Raven, B., "The Bases of Social Power" in: Dorwin Cartwright and Alvin F. Zander (eds.), *Group Dynamic*, Harper and Row, New York, 1960, pp. 607-623.

Geoffrey, T., *Seapower. A Guide for the Twenty-First Century*, Routledge, London and New York, 2nd edition, 2009.

Georges, P., *Le management cockpit. Des tableaux de bord qui vont à l'essentiel*, Editions d'Organisation, Paris, 2002

Gill, P., *Rounding Up the Usual Suspects: Developments in Contemporary Law Enforcement Intelligence*, Aldershot, Ashgate, 2000.

Gladwell, M., *Blink: The Power of Thinking without Thinking*, Little, Brown and Company, New York, 2005, p. 74.

Greenwood, P., Chaiken, J., and Petersilia, J., "The Criminal Investigation Process", Rand Corporation, Lexington, Massachusetts,1977.

Goldblatt, P. and Lewis, C., "Reducing Offending: An Assessment of Research Evidence on Ways of Dealing with Offending Behaviour", Home Office Research Study 187, Home Office, London, 1998.

Hale, C. D., *Police Patrol, Operations and Management*, Pearson Prentice Hall, Upper Saddle River, New Jersey, USA, 2004, chap. 10.

Hastings, M., *Finest Years, Churchill as a Warlord 1940-45*, Harper Press, London, 2009.

Hersey, P. and Blanchard, K. H., *Management of Organizational Behavior*, Prentice Hall, Englewood Cliffs, N.J., 1977.

Heuer, R. J., *Psychology of Intelligence Analysis*, Novinka, New York, 2006, p. 126.

John, T. and Maguire, M., "Rolling Out the National Intelligence Model: Key Challenges" in: K. Bullock and N. Tilley (eds), *Crime Reduction and Problem Oriented Policing*, Willan, Portland, 2003, pp. 38-68.

Leman-Langlois, S. and Dupuis, L., "Les technologies de protection des espaces", in: Cusson, M. B. Dupont and F. Lemieux (eds), *Traité de sécurité intérieure*, Presses polytechniques et universitaires romandes, Lausanne, 2008.

Marx, G. T. and Corbett, R., "Critique: No Soul In The New Machine: Technofallacies In The Electronic Monitoring Movement", *Justice Quarterly*, 8(3), pp. 399-414, 1991.

Mintzberg, H., *Le pouvoir dans les organisations*, Editions d'Organisation, Paris, 2003.

Morel, C., *Les décisions absurdes*, Gallimard, Folios essais, Paris, 2002, chap. 1.

Morris, D. R. and Baudin-O'Hayon, G., "The Strategic Warning for Criminal Intelligence", in: *International Handbook on Risk Analysis and Management*, Center For Security Studies, ETH, Zurich, 2008.

Osterburg, J. W. and Ward, R., *Criminal Investigation, a Method for Reconstructing the Past*, Anderson Publishing, Cincinatti, OH, 1992.

Ratcliffe, J., *Intelligence-Led Policing*, Willan, Cullompton, UK, 2008.

Ribaux, O. and Birrer, S. (2009), "Iterative development of co-operation within an increasingly complex environment. Example of a Swiss regional analysis centre" in: F. Lemieux (eds), *International Police Cooperation. Emerging Issues, Theory and Practice*, Willan, Portland, pp. 81-100.

Schreier, F., *Fighting the Pre-eminent Threats with Intelligence-led-operations*, Occasional Paper No. 16, Geneva Centre for the Democratic Control of Armed Forces, Geneva, 2009.

Skogan, W.G., "The promise of community policing", in D. Weisburd and A.A. Braga (eds), *Policing Innovation: Contrasting Perspectives*, New York: Cambridge University Press, 2006, pp. 27-43.

Skogan, W.G., and Frydl, K. (eds), *Fairness and Effectiveness in Policing*, National Research Council, The National Academies Press, Washington DC, 2003, pp. 248-249.

Swanson, C. R., Territo, L., and Taylor, W R., *Police Administration*, Pearson Education Inc., Upper Saddle River, USA, 2005.

Tufte, R. E., *The Visual Display of Quantitative Information*, Cheshire, Connecticut: Graphics Press, 2001.

Van Creveld, M., *Command in War*, Harvard University Press, Cambridge, Massachusetts, 1985.

Articles

Darling, M., Parry, C., and Moore, J., "Learning in the Thick of it", *Harvard Business Review*, July-August 2005.

Egger, S. A., "A Working Definition of Serial Murder and the Reduction of Linkage Blindness", *Journal of Police Science and Administration*, 12(3) 1984, pp. 348-355.

Ferret, J. "Evaluer la police de proximité?", *Les cahiers de la sécurité intérieure*, 3e trimestre, IHESI, Paris, 2003.

Hammond, John S., Keeney, R. L. and Raiffa, H., "The Hidden Traps in Decision-making", *Harvard Business Review*, September-October 1998.

Kaplan, R. S. and Norton, D. P., "Using the balanced scorecard as a strategic management system", *Harvard Business Review*, January – February, 1996, p. 76.

Rogers P. and Blenko, M., "Who has the D?, How Clear Decision Roles Enhance Organizational Performances", *Harvard Business Review*, January 2006.

Shane, J. M. "Compstat Process", *FBI Law Enforcement Bulletin*, April 2004, pp. 12-20.

Shane, J. M. "Compstat Implementation", *FBI Law Enforcement Bulletin*, June 2004, pp. 13-23.

Vollmer, A. and Schneider, A., "The School for Police as Planned at Berkeley", *Journal of the American Institute of Criminal Law and Criminology*, 6, 1917, pp. 877-898.

List of Tables and Figures

LIST OF TABLES

Chapter 1

Chapter 2

Chapter 3

Chapter 4

Chapter 5

Chapter 6

LIST OF FIGURES

Chapter 1

Chapter 2

Chapter 3

Chapter 4